Get Into Game Dev

Get Into Game Dev: Tech Interview Tactics is a crash-course on how to pass a game development technical interview. It's designed to guide intermediate and experienced coders through the depth and rigor necessary to land some of the most highly sought-after roles within interactive media.

Unlike generic interview-prep books, GIGD maintains a laser-focus on game development to directly prepare candidates for roles like technical designer and gameplay engineer. Topics include 3D maths, programming fundamentals, and software design patterns. The author provides high quality instruction and practice problems based on his experience as a professional instructor and developer.

Key Features:

- Includes an extensive set of practice questions taken from interviews of leading game development studios.
- Synthesizes coding and maths fundamentals into focused instruction, directly applicable to game development.
- Culminates in a rigorous practice test, designed to identify a reader's weaknesses and guide them along the path to mastery.
- Uses a variety of mnemonics to assist readers in memorizing subject matter.
- Provides example worked solutions for readers to compare against their own problem-solving approaches.

This book does not teach game development. Instead, it provides knowledge and instruction for a developer to achieve the technical mastery necessary to become a professional game developer.

Get Into Game Dev

Tech Interview Tactics

Matthew Ventures

CRC Press
Taylor & Francis Group
Boca Raton London New York

CRC Press is an imprint of the
Taylor & Francis Group, an **informa** business

Designed cover image: Gongor G. (@gongor32)

First edition published 2025
by CRC Press
2385 NW Executive Center Drive, Suite 320, Boca Raton FL 33431

and by CRC Press
4 Park Square, Milton Park, Abingdon, Oxon, OX14 4RN

CRC Press is an imprint of Taylor & Francis Group, LLC

© 2025 Matthew Ventures

ISBN: 978-1-032-93362-7 (hbk)
ISBN: 978-1-032-93359-7 (pbk)
ISBN: 978-1-003-56555-0 (ebk)

DOI: 10.1201/9781003565550

Typeset in Times
by SPi Technologies India Pvt Ltd (Straive)

This book is dedicated to the Against Malaria Foundation.

Contents

About the Author

Matthew Ventures is an American game developer with experience on large AAA projects and small, sometimes solo, indie projects. His credits include contributions to the projects Doom: The Dark Ages (2025), God of War: Ragnarök (2022), Call of Duty: Warzone (2020), Tony Hawk's Existential Nightmare (2020), Fortnite Creative (2018), Facebook Spaces (2017), NoStranger (2017), and many more. Ventures is passionate about game development education and has worked with students as an instructor and course designer at Stanford University, as a mentor through organizations like Code 2040 and the IDGA, and as a YouTube content creator.

Technical Reviewers

Sandro Victoria-Arena, Jan Fischer, Alejandro Moya García, and Anonymous. If you identify an error in this book, please notify gamedevbook@gmail.com

Introduction

1

WELCOME

Welcome to *Get into Game Dev: Tech Interview Tactics!*

My name is Matthew Ventures and I'm a professional game developer with a passion for teaching/helping others!

This book is designed for novice developers to help them pass their interviews and for experienced developers who wish to brush up on their skills by reviewing topics or going through the practice questions.

This guide is fool-proof! I would know, because it worked for me, and I am oftentimes foolish. Every time I have failed an interview question, I took detailed notes on the problem and then studied hard to improve my understanding. After years of interviews, this book was formed from a collection of those notes.

In the years this guide has been available, dozens of developers have reached out to me over email and LinkedIn to share how it helped them. Check out some of their stories in the testimonials section! I'm confident that if you're willing to put in the time studying, this guide can help you as well.

Let's get started!

TESTIMONIALS

I just wanted to follow up and thank you for your advice and all the time you put into helping students break into the game industry. Thanks to your help and advice, I was able to land an internship as a gameplay engineer at Blizzard on the Diablo IV team.
— Zakarya Zahhal, 2023

DOI: 10.1201/9781003565550-1

I recently experienced a studio closure and had to start the hunt for a new job with no warning. *'Get into Game Dev'* by Matthew Ventures proved invaluable in my job hunt. Even as a senior engineer with nearly a decade in the game industry, this guide helped refresh my knowledge, especially when it came to C++ and 3D math. Both those new to game development and more experienced folk can get a lot out of it!

—Thomas Klovert, 2023

Your guide was invaluable! All of the questions during my final interviews were straight from your guide! It was unbelievably helpful! I would absolutely recommend it to any other aspiring game devs.

—Avik Shenoy, 2023

I just wanted to say I super appreciate all the work you've put into the *Get Into Game Dev* book and your YouTube. I've been programming for a long time now, but self-taught, and there's a lot of things I can do but wouldn't use the right terminology when discussing with other coders. This seems like a great way for me to work on that :) Thanks again!

—Ian Snyder, 2023

11 out of 10. It taught me information not explicitly told in my programming classes and provided me with incredibly valuable insight.

—Frazier Kyle, 2024

I wanted to let you know that your course helped me with the id Software UI Programming interview. I made it through all of the rounds including the 'onsite'.

—Zak Cook, 2024

Recently bought your book and it's been incredibly helpful so far. I'm still in the process of putting a portfolio/resume together, but once I (hopefully) get an interview, I know I'll be very prepared. :)

—John Isril, 2024

WHO IS THIS BOOK FOR?

This book is written with a candidate gameplay engineer in mind. A gameplay engineer (sometimes referred to as a GPE) is a coder on a game development team who has a strong depth of competency when it comes to programming (and maths!) but who also has a breadth of knowledge in adjacent disciplines such as design and animation. The GPE is, in many ways, one of the most generalist roles in game development. So, there's a lot of material for us to cover! But I would encourage you to also extend your studies based on the exact needs of the role and company you are applying to. For example, if you're applying to a graphics programmer role, you should invest additional time into graphics programming; and if you're applying to a team with many networked multiplayer games, then it would be wise to self-study additional materials focused specifically on networked programming.

HOW TO READ THIS BOOK

Start with the end. Jump ahead to the final chapter and give yourself focused time to think through the questions and try to solve them on your own. Based on your performance, jump to chapters that cover subject areas where you feel you could improve. Skim chapters until you reach the practice questions and if you're not sure how to approach them, read the preceding paragraphs until you do. As you work through the book, I encourage you to research topics on the internet for additional learning material. The subject matter of game development, even when focused on gameplay engineering, is simply so vast that there is no way this book will be able to exhaustively teach every trick of the trade. Instead, the greatest value of this book is to provide you with the perspective to realize what things you don't yet know and to empower you to learn those things through self-study and experimentation.

As you'll see, many interview questions will ask you to select an approach from several competing options. Interviewers want to understand how your knowledge and reasoning skills are used together. Whenever you come across a new concept, you should ask yourself how it relates to what you already know, and you should anticipate questions like "what are the benefits and drawbacks to this approach that could inform when to employ it?". If you don't feel confident answering those questions, I encourage you to plug them into ChatGPT or Google in search of answers.

This book frequently assumes that you are familiar with the basics of programming and maths. For example, it doesn't invest time to explain what a pointer is before diving immediately into the tradeoffs between different kinds of pointers. I've done my best to keep the book focused and trim. If you encounter a term or topic you've never come across before, I'd recommend taking a moment to self-study the topic before you resume reading. To some extent, that's the main purpose of this whole book: helping candidates realize topics they're not familiar with.

PREPARING FOR A TECHNICAL INTERVIEW

While in some cases you may be invited to a direct interview with a company, most candidates must first pass one or more preliminary interview stages. This book does not include detailed preparation steps for the pre-interview, but I nonetheless wanted to provide you with some high-level guidance on those steps in the process. Prior to a technical interview, candidates will typically need to provide portfolio projects and pass a phone screen.

Portfolio Projects

Most candidates will begin their application by first completing portfolio projects to present to a company. These projects are often misunderstood as a measure of a candidate's creativity. Portfolios for technical roles are meant to demonstrate technical competency. While it's certainly impressive if a candidate develops and publishes their own indie game, generally that level of scope is not necessary for a project.

A great project can be as simple as implementing the technical requirements of a single gameplay feature. For example, while I was working at Sony Santa Monica, I saw several candidates submit their implementations of the famous Kratos axe throw from *God of War* (2016). Recreating an iconic feature from a game made by the studio you are applying to is a great idea! It's somewhat flattering and helps the team think: "we should hire this person; they could have implemented that feature for us!".

FIGURE 1.1   An example portfolio project of Kratos' axe throw [1].

You'll know if a project is sufficient to be a portfolio piece if you faced significant technical challenges during its development. In an interview, you'll be asked to describe those challenges and how you overcame them. The important thing is that the project was completed and that the skills it required are relevant to the role. For example, it would be inappropriate to apply to a programmer role with portfolio pieces solely implemented using scripting, such as through Unreal Engine's "Blueprints" (a visual scripting language). Programmers should be programming, which means using whatever language is used by programmers at the company you are applying to. Most companies have programmers code and designers script.

Portfolio projects are usually presented asynchronously, by sending a hiring manager a website that includes media demonstrating the projects. It's important that you surface the impressive parts of your project, making them as easy to access and evaluate as possible. Do not ask a hiring manager to "download this game and play until level five to see the cool boss battle". Typically, the projects are presented in a video format, with the visuals showing aspects of the project and the candidate providing voice-over to contextualize and explain each aspect as it comes up. Some students make the mistake of just using their game project's trailer as the media piece. It's not a trailer, you're not trying to get the hiring manager to buy or even play the game. The video should instead be more of a behind-the-scenes explainer that details how a specific feature of your project was implemented.

The most common mistake I have seen at this stage is a lack of focus. Many candidates, students in particular, have worked across disciplines and will confidently call themselves artists, designers, modelers, animators, musicians, and programmers. While this is certainly impressive, it is often sadly also irrelevant and may actually hurt a portfolio. Unless the role you are applying to explicitly requires competency across each discipline, you should probably focus a single portfolio on your chosen discipline. And when developing a portfolio project, if you need an asset from another discipline, you should almost always elect to purchase or download a pre-existing asset that was developed by an expert. Just make sure to clearly credit other people's work on the project.

Resumes

A resume is a succinct summary of your prior, relevant, work. If your prior experience is mostly making portfolio projects, find a way to present them in your resume as work experience. If you self-published a game, then congratulations you're an entrepreneur. Give your "company" a name and slap that project into your resume as work experience. When writing a resume, always focus on your achievements rather than responsibilities. Demonstrate your work's impact through quantitative data as much as possible. Ideally, each bullet point reads like a formula: action + skill = result. Here's an example: "Led feature development to support limited-timed discounts in the game's store, using C++ and Blueprint scripting, to increase in-app-purchasing spend by 10%." Never quantify your skill competency, for example never say, "I have 75% C++ skill". Instead show your skills by including them in your bullet points. Use a varied set of strong "action verbs", MIT has published a fantastic set for you to reference [2].

Phone Screens

A phone screen is a conversation you'll have with a recruiter or hiring manager prior to the technical interview. Sometimes a phone screen will contain technical questions, but typically it is just designed to gage if you are worth an investment of the team's time. Some smaller game development studios will completely outsource this step of the process, in which case you may need to solve a basic coding problem on a website like LeetCode. During a phone screen, your objective is to express a great amount of interest in the studio and strong competencies aligned with the needs of the role.

Studios usually request critique of their work in order to measure your familiarity with their projects and the broader landscape of their genre. I always recommend candidates play the latest game from the studio and be prepared to answer critical questions about the game. For example, many studios will ask "how could the game be improved?" I advise you to prepare your answers to questions of this nature by focusing on what was great and could be further expanded rather than on what was not-so-great and could be revised or cut. Certainly, there are many games where unnecessary elements negatively impacted the game experience. But strategically, it's always better to stay focused on the positives rather than the negatives, even if you can eloquently explain them. You never know if the feature you identified as lackluster may have been developed by the person conducting the interview! It's best to always consider the glass "half full". I've only failed one phone screen…I had the privilege to interview with the Fallout 76 team of Bethesda Games Studios Austin. When asked about what I disliked about the game, I explained that the quality level was below a bar I would feel comfortable putting my name on and then I explained what methodologies I would use to approach reducing technical bugs and regaining consumer trust. This was not a "correct" answer. Please learn from my mistakes. I should have said something like "the quest system was awesome and provided me endless gameplay, and I think it could have been further improved by adding faction reputation so that factions could react-to and reward me based on how much I helped them". Notice how that answer basically had nothing negative. If you are familiar with improvisational acting, the principle at play here is to take the "yes and" approach, in order to build upon their work.

If you're asked about "what games are you playing" or "what's your current favorite game" you'll want to steer your answer toward games that are recent, critically praised, commercially successful, and in the same genre as the games of the company you are applying to. For example, you certainly do not want to tell the Call of Duty recruiter that your life is dedicated to League of Legends because that will only lead them to wonder "why are they applying here and not Riot Games?" As an interviewer for id Software, I met with a surprising number of candidates who expressed zero interest in first-person-shooters (the only genre id Software has ever worked in). While that may be okay for certain roles, a gameplay engineer needs to understand the game feel of the genre they are working in. When you choose a game, be prepared to discuss what you liked and disliked about it (keeping in mind the "yes and" approach described previously). I encourage you to reference critiques of the games you plan to discuss. Searching "game critique [game name]" on YouTube is a great place to start.

At this early stage, your interviewer may ask you for salary expectations. It's almost always in your best interest to delay this negotiation as far into the process as possible. The compensation negotiation is a topic for another book, but the general strategy is to delay the conversation until the team is maximally invested in your candidacy. Wait until you've used the skills you've learned from this book to ace their interview before providing your compensation expectations. And to the extent that you can, try to inform your expectations based on as much data as you can collect. I recommend the websites Levels.fyi and Glassdoor as sources for researching salaries (though many game studios do not appear in their listings).

WORKS CITED

[1] A. Cadogan, "[UE4] God of War Axe Throwing," *YouTube*, Aug. 2024. [Online]. Available: https://www.youtube.com/watch?v=nWIxbafZ1hw&ab_channel=AidanCadogan. [Accessed: Aug. 16, 2024].

[2] "Resume Action Verbs," *Center for Academic and Professional Development*, MIT, [Online]. Available: https://capd.mit.edu/resources/resume-action-verbs/. [Accessed: Aug. 26, 2024].

PART I

Programming

Fundamentals

2

BITWISE OPERATORS

It seemed to me that the best way to start this guide would be with the very building blocks of all data: bits. Bits are our smallest unit of data and are used to build the larger data types we will discuss later on. A bit is a single number: 1 or 0, in other words, ON or OFF. Usually when we need a two-state variable we will use a bool (more about bools later) but they are 8 bits! So, to save space, it is common to instead split our bits so that each bit in an 8-bit structure can represent a different parameter's on/off state. When we treat each bit in a data type as its own value, that's called a bit field. Now, instead of only two states, we can easily represent 8*2=16 states within the same 8-bit structure! And if we consider several bits within a bit field instead of just one at a time (a process called bit-masking) we can represent many more combinations.

To manipulate a single bit within a bit field, we can use the bit-wise operators. Here's an example using the AND operator on two ON bit operands which produces an ON bit output.

1 & 1 = 1

Here are some operators you should be familiar with:

- &, called "AND", is ON only if both bits are ON (1&1=1, 1&0=0)
- |, called "OR", is ON if either of the two bits is ON (1|1=0, 1|0=1)
- ~, called "NOT", inverts the value of a bit (~1=0, ~0=1)
- ^, called "XOR", is ON if the bits are different. Another way to think of this is that it outputs ON if either one of the input bits is on, but not both. (1^1=0, 1^0=1)
- <<, called "left shift", moves bits of the 1st operand to the left by a number of places specified by the 2nd operand. (00010010 << 2 = 01001000)
- >>, called "right shift", moves bits of the 1st operand to the right by a number of places specified by the 2nd operand. (11010011 >> 2 = 00110100)

DOI: 10.1201/9781003565550-3

These are your tools…

&	|	~	^	<<	>>

… now let's discuss some ways to use them!

BITWISE OPERATIONS

Below are some common bitwise operations. Before reading their descriptions, I encourage you to quiz yourself to see if you can think through how to achieve each operation using our bitwise operators.

Set the Nth Bit

When we say "set" a bit, we always mean setting the bit to ON, meaning 1, since that is the only non-zero value a bit can have. Bits are numbered from right-to-left, so if you are asked to set the zeroth bit, that refers to the bit on the far right which is sometimes called the least significant bit (LSB).

myVar = myVar | (1 << n)

In the above implementation, we are taking any input number (represented as "myVar") and setting it to be equal to itself OR'd with the result of a leftward bit shift. Let's breakdown each step in the process:

1. First, we create the number 1 which is represented in bits as 0001 (though the exact number of preceding zeroes may vary). All of the bits in our number are now 0 except for one which we will call our "special" bit.
2. Then we shift the bits to the left by n, which will push our special bit into the nth position. For example, if n==0 and we are trying to set the zeroth bit, then 1<<0 will output the bits 0001 with our special bit still in the LSB position. And if n==2 and we are trying to set the second bit, then 1<<2 will change the bits to 0100 with our special bit now in the "second" position (which is the third from the right because we start counting at zero).
3. Then we use an OR which will do an OR operation between each bit of the original number and the corresponding bit in our shifted result. The OR operations that include 0's from our shifted result will leave the bits of the original number unchanged. But the single OR operation that corresponds to our special bit will also result in an ON bit, because our special bit is 1 and no matter if the other bit is 0 or 1, the output will be 1 (0|1=1 and 1|1=1).
4. Our final step is to assign the result to myVar. This step is usually combined with the previous step using the |= operator which means "do an OR between

both operands, and then assign the result to the left operand". So, while in this case I wrote out this example as "myVar = myVar | 1 << n", you'll usually see this operation written as "myVar |= 1 << n".

Here's what those steps look like in an example where we want to set the third bit of myVar which currently holds the number 0100.

0. myVar = myVar | (1 << n)
1. myVar = 0100 | (0001 << 3)
2. myVar = 0100 | 1000
3. myVar = 1100
4. myVar is now 1100

The important strategy to take away from this approach, is that by using the number "1", we can build a new number to affect the exact bit we want to change. I should clarify that we always use an unsigned value for this purpose (to prevent numbers becoming negative, a concept we will discuss later). I won't go into the same detail for each of the following operations, but this general technique is used repeatedly (basically anytime we introduce a 1). Again, I remind you to challenge yourself and test if you can figure out how to complete each of the following operations without looking at the implementation. Don't be afraid to work through the problem by writing down the bits at each step of the process.

Clear the Nth Bit

Clearing a bit means setting its value to OFF.

myVar & = ~ (1 << n)

Toggle the Nth Bit

Toggling a bit means setting its value to be the opposite of its current value.

myVar ^ = 1 << n

Check the Nth Bit

Checking a bit means determining whether it is ON or OFF. This is done by returning a value of zero if the bit is OFF or returning any non-zero value, typically 1, if it is ON.

output = (myVar >> n) & 1

Check Bitflags and Enumerations

As mentioned earlier, you could design a program where each bit of a number represents a variable. For instance, with three bits you could have the first bit represent if a character can fly, the second bit represent if they can swim, and the third represent if they can teleport. Below are two examples where the first (leftmost) three bits are used to represent traversal capabilities.

```
Traversal Capability:
```
- 10000000 = character can fly, but not swim or teleport.
- 01100000 = character cannot fly, but they can swim and teleport.

The bitfields used in this approach are called bit flags. Each bit is like a literal flag that can be raised or lowered and represents one of two states, independent of the other bits.

Another approach, to represent data in bitfields, considers each unique combination of bits to be a separate value, called an enumeration ("enum" in code). This approach allows you to represent many more values using the same number of bits. For example, here are five different values we can represent as enumerations, using only the fourth through sixth bits from the left.

```
Monster Type:
```
- 00000000 = zombie
- 00010000 = vampire
- 00001000 = werewolf
- 00000100 = mummy
- 00011100 = ghost

Let's walk through an example wherein a character is represented with eight bits. From left to right, the first three bits represent a character's traversal capability (using bitflags as explained previously), the next three represent a character's monster type (using enumerations as explained previously), and the final two bits are unused.

Check Bitflags

```
Character: 10100100
```

First let's test if this character can fly and swim. To do this need to create a helper bitfield called a mask to represent the bitflags required to satisfy this check.

```
flySwimMask = Fly | Swim
flySwimMask = 10000000 | 01000000
flySwimMask = 11000000
```

Now we check if the value AND'd with the mask produces the mask itself.

```
output = ((myVar & MASK) == MASK)
```

```
output = ((Character & flySwimMask) == flySwimMask)
output = ((10100111 & 11000000) == 11000000)
output = ((10000000) == 11000000)
output = False
```

The result is false. Even though this character can fly, they cannot fly <u>and</u> swim.

Check Enumerations

```
Character: 10100100
```

Using the same character let's now check if they are a werewolf. To do this, we again create a mask, this time to isolate all of the bits that are used to represent the monster type.

```
MonsterTypeMask = 00011100
```

Now we check if the input value AND'd with the mask is equal to our expected value.

```
output = ((myVar & MASK) == EXPECTED)
output = ((Character & MonsterTypeMask) == werewolf)
output = ((10100100 & 00011100) == 00001000)
output = (00000100 == 00001000)
output = False
```

Practice Problem 2.1: Complete implementations for the following functions...

```
bool HasAnyFlagSet ( int inputFlags, int flagsToCheck ) { }
bool HasAllFlagsSet ( int inputFlags, int flagsToCheck ) { }
int EnableFlags ( int startingFlags, int flagsToEnable ) { }
int DisableFlags ( int startingFlags, int flagsToDisable ) { }
```

Solution:
```
bool HasAnyFlagSet ( int inputFlags, int flagsToCheck ) {
      return ( inputFlags & flagsToCheck ) != 0;
}
bool HasAllFlagsSet ( int inputFlags, int flagsToCheck ) {
      return ( inputFlags & flagsToCheck ) == flagsToCheck;
}
int EnableFlags ( int startingFlags, int flagsToEnable ) {
      return ( startingFlags | flagsToEnable );
}
int DisableFlags ( int startingFlags, int flagsToDisable ) {
      return ( startingFlags & ~flagsToDisable );
}
```

BINARY AND HEXADECIMAL NUMBERS

The bits we discussed so far have two states ON or OFF, 1 or 0. Binary (where "bi" means two) is the representation of a number using only these two values: 1's and 0's. But you are probably more familiar with numbers in decimal format, a format where each digit has ten states (0-9). Binary numbers are just another way of representing decimal numbers and they are specifically an easier format for the computer (which is composed of binary switches). Another common format is Hexadecimal (16 states), its compactness makes it easier for humans to read. As a video game programmer, you should be able to translate numbers between these formats. It's important to understand the implications of these representations because they directly determine how much data a variable can hold. For example, if each digit of a number can only be one of two values, then the total amount of permutations that number can represent is calculated by 2^n where n is the number of digits.

$$2^0 = 1 \bigm| 2^1 = 2 \bigm| 2^2 = 4 \bigm| 2^3 = 8 \bigm| 2^4 = 16 \bigm| 2^5 = 32 \bigm| 2^6 = 64 \bigm| 2^7 = 128 \bigm| 2^8 = 256$$

Companies (particularly Sony studios) will sometimes ask candidates to manually compute powers of two. As a mnemonic, I suggest memorizing "2 to the 6 is 64" by singing it aloud like a little jingle. The alliteration of "six" and "sixty-four" makes it a little easier to remember. Memorizing that relationship can help you jump to adjacent powers such as 2^5 with a simple 64/2=32.

Practice Problem 2.2: What is 2^8 and what is the significance of this value?

> **Solution**: 2^8 = 256 which represents the number of unique permutations we can create with an eight-digit binary representation of a number. This is why an unsigned int-8's maximum value is 255 (it would be 256 if we did not also need to represent zero).

Practice Problem 2.3: What is the maximum value of a signed int-32?

> **Solution**: An unsigned int-32 would have all 32 bits to represent its value, but we need to remember to subtract one because one combination of bits is used to represent 0. Therefore, the max value of a uint-32 is $2^{32}-1$. But I asked for a signed int-32 which requires the first bit to represent the sign. Therefore, the max value of a singed int-32 is $2^{31}-1$, or 2,147,483,647.

You may have noticed that in Practice Problem 2.3, I tried to trick you by asking for the signed max instead of the unsigned max. Since we are representing a signed number, we need to split all possible permutations of our 32 bits into three categories: positives, negatives, and zero. With three categories we cannot evenly divide the permutations. The result is that the absolute value of Int32.MinValue is not equal to the absolute value of Int32.MaxValue. Hopefully that provides you with a hint for the next practice problem.

Practice Problem 2.4: What is the minimum value of a signed int-32?

Solution: We learned that the maximum value of a signed int-32 is $2^{31}-1$. That's because of the 32 bits, one needs to be used for the sign (hence the exponent of 31); and because one combination of bits is used to represent zero (hence the -1). Because we accounted for zero by lowering the maximum value by one, we do not need to constrain the minimum value. Therefore, the minimum value of a signed int-32 is simply -2^{31}.

POSITIVE NUMBERS IN BINARY

The first step to understanding binary representation is realizing that, even in decimal form, we break down a number by sorting its digits into columns. For example, consider the number 142, we call this number "one hundred and forty-two". We can be explicit about the value of each column by saying: "one hundred, four tens, and two ones". But another way to think of these columns is to consider each as a power of ten: $10^2=100$, $10^1=10$, $10^0=1$. So, in order to get your brain ready for binary numbers, try to think of the ones column as the 10^0 column, the tens column as the 10^1 column, the hundreds column as the 10^2 column, and so on. Because each column label is formed by an exponential expression where ten is the base, we call this system "decimal" (where "dec" means ten).

1 4 2

$$1\times10^2 + 4\times10^1 + 2\times10^0$$

FIGURE 2.1   142 expressed as $1 \times 10^2 + 4 \times 10^1 + 2 \times 10^0$.

Binary representation uses the same system of exponents and columns, the only difference is that the base is 2 instead of 10. For example, consider the number 1011 in binary. Working through the columns from left to right, this number is $(1 \times 2^3) + (0 \times 2^2) + (1 \times 2^1) + (1 \times 2^0)$. You can work out the math to convert this number into the decimal representation of 11.

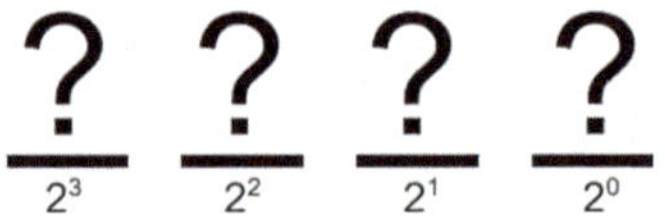

FIGURE 2.2   Four columns, each representing a power of two.

When converting numbers from decimal to binary, I typically use a computer! But if I had to do it manually, I would start from left to right filling in columns as I go. You can imagine that each column is a question mark as shown in Figure 2.2 and you are determining if the value of each column is 0 or 1. Let's work through converting the number 10 from decimal into binary. We start with the left-most column which is $2^3=8$. For this column, we ask, "Is our current number of 10 greater than or equal to 8?" Since it is greater, we put a 1 in that column. Since that column represents a portion of our number, we only need to worry about the remaining amount, so we calculate 10−8=2. And now 2 is the number we use to consider if a digit should be placed into the next column. For the next column, 2 is less than 2^2, so we should leave a zero in that column. The third column from the left is 2^1 which is equal to 2 so we mark a 1 there. We have no remaining value to represent, so any remaining columns to the right (in this case there is only one) should be filled-in with 0. The result is 1010 which is how we write the number ten in binary.

POSITIVE NUMBERS IN HEXADECIMAL

Hexadecimal numbers, or "hex" numbers, work just like binary numbers except the base is 16 instead of 2. And while binary digits can only be 0's and 1's, hex digits can be 0 to 15. Because we can't write numbers above 9 as a single digit, we switch to letters once we get to 10. Here's a simple table to help you convert between the two systems.

Decimal	0	1	2	3	4	5	6	7	8	9	10	11	12	13	14	15
Hexadecimal	0	1	2	3	4	5	6	7	8	9	A	B	C	D	E	F

NEGATIVE NUMBERS IN BINARY

In an interview, you may be asked to represent a negative number. Interviewers like this question because it also requires you to understand the positive representation. In total, the three steps to writing a negative number in binary or hex are:

1. Write the number as positive
2. Take the ones' complement of the number
3. Add one

Step two mentioned the ones' complement which is a straightforward way of saying "the number I need to add to a binary number to make it 'all ones.'" For example, if we take the binary number 1001, we can see there is a two-digit gap in the middle. The number I need to add to 1001 so that the sum is all ones is 0110, which will fill the gap. Therefore 0110 is the ones' complement of 1001. In practice, for binary numbers, this is the same as simply flipping the bits (1's become 0's and 0's become 1's). But finding the ones' complement of a hex number requires us to abstract the process a bit. A more generalized way to think about it is subtracting a column's current value from the max value that column can represent.

Max – Cur = ones' complement

For example, if we have a hex column with the value A, we subtract it from the maximum value that a hex column can represent, F. The result: F–A = 15–10 = 5. This approach will work to find the ones' complement for any number representation, whether it be binary, hex, octal, etc. For example, if I want to take the ones' complement of 54 in octal, I'll plug each digit into the formula and solve. Remember, we do the formula for each column independently.

Ones' complement of 54 in octal:

For the 5: Max-Cur = 7–5 = 2
For the 4: Max-Cur = 7–4 = 3

Therefore, the result for 54 is 23.

If I was writing this book in the 1980's, we could stop here because computers in those days simply used a number's ones' complement as its negative representation. But ones' complement has an issue which is that there are technically two ways to represent zero (a "negative" zero and a "positive" zero). So modern computers add a final step to convert a number to its two's complement instead.

After determining a number's ones' complement, the final step to making it negative is to simply add one. This final step forms a number known as the two's complement which is finally a number's negative representation. Let's work through some examples:

Practice Problem 2.5: How would you represent -5 in binary using eight bits?

Solution:

Step 1: Write 5
0000 0101

Step 2: Take the ones' complement
1111 1010

Step 3: Add 1
1111 1011

Practice Problem 2.6: How would you represent -2 in hexadecimal using eight bits?

Solution:

Step 1: Write 2
02

Step 2: Take the ones' complement
FD

Step 3: Add 1
FE

Did you notice that the hex number in Practice Problem 2.6 used only two digits, "02" to represent eight bits? That's because hex numbers are implemented just like binary numbers under-the-hood. Each hex digit represents four binary bits. So "02" in hex is the same as 00000010 in binary. Sometimes, when written out, the following notation is used:

0b10000 is sixteen in binary, the "b" denotes binary.
0x00000010 is 16 in hex, the "x" denotes hex.

Rarely in interviews, candidates will be asked to manually solve subtraction and additional problems using hexadecimal numbers. While addition is straightforward, subtraction is not. I recommend you solve subtraction by first making the subtracted term into a negative number and then you can simply add the negative number to the number you are subtracting from.

Practice Problem 2.7: How would you represent -42 in hexadecimal using eight bits?

Solution:

Step 1: Write 42 in hex.
I know that my answer will be a number with several columns. I start by writing out labels for my columns from right to left to assess how many

columns I will need. First, $16^0 = 0$, then $16^1 = 16$. These two columns are respectively the 0's column and the 16's column. The highest value I can place a hexadecimal column is 15 (F), and a 15 in the 16's column would represent $16^1 * 15$ which is a value far beyond 42. This tells me that I will not need a column larger than the 16's column in order to represent 42.

To determine which number to put in the 16's column, I divide 42 by 16. The result is 2, with 10 remaining. Therefore, I place a 2 in the 16's column and bring the remainder of 10 into the 0's column. The 0's column is our final column and we place the entire remainder of 10 inside by using the letter A. Remember that each column can only have one character, so when we need to represent a two digit number, it is written as a letter (in this case 10 becomes A).

Our final answer for this step is therefore 2A.

Step 2: Take the ones' compliment.

In this case we are working in hex, Let's put both column values, 2 and A, through the formula.

$$2 : 15 - 2 = 13 = D$$

$$A : 15 - A = 15 - 10 = 5$$

Our final answer for this step is therefore D5.

Step 3: Add 1

In this step we add 1 to the entire number (not each column). Our 0's column is currently 5, 5+1=6. 6 does not exceed 15 (the maximum value we can hold in any particular column) so we don't need to worry about carrying over any values to the 16's column.

Our final answer for this step is therefore D6.

In hex, D6 represents 8-bits. If we were instead using a 16-bit representation, then we would have 00D5 as our answer to Step 1, and FFD6 as our answer to Step 3.

FLOATING-POINT NUMBERS

We just explored how fixed-point numbers are represented at the bit level. Fixed-point is a great representation for whole numbers, but when we need to express decimal numbers, we typically use floating-point. Floating-point numbers or "floats" have an interesting tradeoff: they can express very large numbers (both negative and positive) but as the magnitude of the numbers increases, the precision of a float's representation decreases. To understand this phenomenon, let's first take a look at how a float is laid out in memory:

1   **01110110**   **01001111011011100011101**
1-bit sign   8-bit exponent   23-bit significand

FIGURE 2.3   A float divided into three parts: a 1-bit sign (if this bit is ON, then the float's value is negative), an 8-bit exponent (from which we subtract a bias of 127), and a 23-bit significand aka mantissa (this is a "normal" (fixed-point) binary number).

An interesting decision was made here. The float's designers chose for the exponent bits to represent a number from which we would subtract 127 instead of representing the exponent itself. That allows for fractions because 127 subtracted from a small number will yield a negative exponent which produces a fraction.

A float's lack of precision at large magnitudes is its general downside. This was a big problem for Minecraft and resulted in odd behavior when positions were far enough away from the world origin for the terrain generation to no longer function properly. This resulted in strange uniform cliff faces known by the community as the "Far Lands". In one studio I worked at, the team resolved this issue by introducing intermediary transforms for each section of the level, so that we only needed to make calculations relative to a somewhat-nearby origin (that way the numbers never got too big). In Star Citizen, the team used double-precision floating-point values called "double floats" which were implemented like floats but just double the size. I am including these stories because it is important to remember that we are not simply exploring "math land", all of this knowledge has a grounded purpose and direct application to game development.

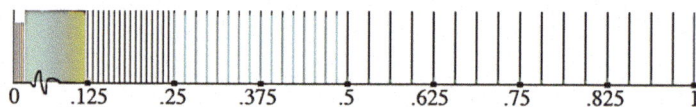

FIGURE 2.4   The values a float can represent are crowded near zero and more spread out as the values grow in magnitude [1].

Floats lose precision at higher magnitudes because the set of values that they can represent is not uniformly spread out across a number line. Their set of possible values is extremely dense around zero and then increasingly sparse as magnitude increases. Numbers too close to zero to be represented are called "subnormal" (and occasionally called "denormal"). How can we deal with these limitations?

If you are dealing with small and big numbers, make sure to deal with your big numbers first. For example, try to determine which of these will be more accurate (assume positive float values).

$$(\text{big} + \text{small}) - \text{big} \ \textbf{OR} \ (\text{big} - \text{big}) + \text{small}$$

In theory, these expressions yield the same result. But in practice, the second option is more accurate because, by subtracting the big numbers first, we will have a smaller operand to add to our small number. If you have a very big number and a very small number adding them may not change the big number at all! So, we want to avoid that approach because it will provide a less accurate result.

Ultimately, we can't avoid float inaccuracy. Instead, we need to work around it. This includes having a tolerance threshold when checking if numbers are "equal" or "zero". Coders will usually implement functions like IsNearlyEqual and IsNearlyZero to check if a float's value is close enough to the comparison value.

Practice Problem 2.8: Implement the IsNearlyEqual function.

> **Solution**: Nothing too fancy here, our solution is to simply check if the difference of the values is below an acceptable threshold which will vary based on the use case. Sometimes this threshold is accepted as a third argument.
>
> ```
> bool IsNearlyEqual( float a, float b ) {
>     return abs(a - b) < 0.001f;
> }
> ```

DATA TYPE SIZES

Before we get started talking about data types, you should know the following conversions.

- 1 B (a byte) = 8 bits
- 1 KiB = 1024 B
- 1 MiB = 1024 KiB
- 1 GiB = 1024 MiB
- 1 TiB = 1024 GiB

The terms that include an "i" such as "KiB", are called binary prefixes and are used to differentiate from decimal prefixes like KB and GB. The right-hand side of each binary prefix expression is 1024 because that's the value of 2^{10}. Whereas a decimal prefix would be a ratio of 1:1000, for example 1 KB = 1000 B.

If you'd like a mnemonic for remembering the order, I came up with "Burger King Makes Great Toast". Additionally, I like to think of a Mega-Byte (MB) as a "Million"-Byte. I haven't come up with a great way to memorize that there are eight bits in a byte, I typically just tell my students to imagine "byte-ing" off their thumbs, which would leave them with eight fingers (but they think that's weird).

Let's now look at some common data types to understand the amount of memory each one needs. Note that some data types have a different size based on if the system is 32-bit or 64-bit, but pretty much all modern game development is 64-bit.

DATA TYPE	32-BIT SIZE	64-BIT SIZE (IF DIFFERENT)	REPRESENTATION
Bool	1 byte (8 bits)	-	-
Char	1 byte (8 bits)	-	-
Short	2 bytes (16 bits)	-	-
Int	4 bytes (32 bits)	-	Two's Complement
Long	4 bytes (32 bits)	8 bytes (64 bits)	Two's Complement
Float	4 bytes (32 bits)	-	Floating Point
Double	8 bytes (64 bits)	-	Floating Point
Pointer	4 bytes (32 bits)	8 bytes (64 bits)	-

While an int is typically 32 bits, we often need to represent a number with a much smaller range and don't need all of those bits. For these cases we can save memory by using an int type with an explicit size such as an int8_t which only uses 8 bits.

Since pointers are used to represent memory addresses, the pointer's size (and therefore it's maximum value) implicitly defines how much memory a program can use. A 32-bit pointer can only access values up to about 3.5 GB of RAM (this is a serious limitation), whereas a 64-bit pointer could theoretically access values up to 16 billion GB of RAM.

Below are some game file sizes for perspective. Halo 1 (Combat Evolved) was released on the original Xbox (which was 32-bit) the later entries were released on 64-bit hardware.

Halo 1 (2001) - 1.2 GB
Halo 2 (2004) - 20 GB
Halo 3 (2007) - 55 GB
Halo 4 (2012) - 55 GB
Halo 5 (2015) - 95 GB
Halo Infinite (2021) - 48 GB

Practice Problem 2.9: How large is an int* on a 64-bit system? What about a float* on a 64-bit system?

Solution: Both int* and float* are pointers so they will both be the size of a pointer. The int vs float distinction tells us what kind of data the pointer is pointing to, but it's irrelevant to this discussion. A pointer is eight bytes (64 bits) on a 64-bit system, so both will be eight bytes.

Practice Problem 2.9 mentioned a "64-bit system" which refers to a processor's register size and usually specifies something called the "word size" as well. A word is usually the amount of data that can be manipulated at once by a processor. Pretty much all modern processors are 64-bit meaning that the processor can (typically) only work on 64 bits (aka eight bytes) of data at a time. You may come across the terms halfword and doubleword (aka DWORD) which are simply multiples of that value.

So, if someone mentions an n-bit architecture, the word size is probably n bits.

DATA TYPE	32-BIT SIZE	64-BIT SIZE
WORD	4 bytes (32 bits)	8 bytes (64 bits)
DWORD (double word)	8 bytes (64 bits)	16 bytes (128 bits)
QWORD (quadruple word)	16 bytes (128 bits)	32 bytes (256 bits)

ALIGNMENT AND PADDING

With some rare exceptions, the size of a data type also determines an attribute called its "alignment". For example, an int is 4 bytes large and therefore 4-byte aligned. When data types are stored into memory they are stored at the next available alignment boundary of appropriate size. Let's break that process down with an example.

```
Data:     X1 |    | Y1 | Y2 | X2 |    |    | X3 |    |
Address:  0 | 1 | 2 | 3 | 4 | 5 | 6 | 7 | 8 | 9 | 10
```

FIGURE 2.5 Data addresses where indices 0, 4, and 8 are marked as the 4-byte alignment boundaries.

In Figure 2.5, I marked the 4-byte alignment boundaries with X1, X2, and X3. If we ask the compiler to store an int at Y1, it will observe that this address does not correspond with a 4-byte alignment boundary, and it will instead add bytes of padding to positions Y1 and Y2. Then it will place the int at position X2, so that it is aligned. In simple terms, the compiler always finds the next aligned address; and in the case of 4-byte alignment, those addresses are ones with values that are a multiple of four.

Aligning data can drastically decrease the time it takes to access it, and some processors simply cannot perform reads on unaligned addresses. In C++, the compiler always tries to automatically align data which can lead to some interesting behavior. For example, consider the following UncompiledData struct:

```
struct UncompiledData {
    char Data1;
    short Data2;
    int Data3;
    char Data4;
};
```

This UncompiledData struct is eight bytes large prior to compilation:

$$1 \text{ byte}(\text{char}) + 2 \text{ byte}(\text{short}) + 4 \text{ byte}(\text{int}) + 1 \text{ byte}(\text{char}) = 8 \text{ bytes}$$

However, after compilation, the compiler will add padding to align the data. The resulting struct will look like the following CompiledData struct:

```
struct CompiledData{
    char Data1;
    char Padding1[1];
    short Data2;
    int Data3;
    char Data4;
    char Padding2[3];
};
```

The compiler added a total of four bytes of padding to align the data members. This means that, at runtime, the struct will actually use twelve bytes of memory instead of eight!

You may be wondering, "why do three bytes of padding follow the final data member"? Structs do this so that if there are multiple of them (such as in an array) then each will be aligned. Most hardware is specifically built for 4-byte alignment and the compiler will pad any struct until it is aligned (unless you explicitly tell the compiler not to). But 4-byte alignment is only the minimum, if your struct contains a member type that is aligned to a larger byte count, then the compiler will align based on that member instead. For example, if we change Data3 from an int to a uint64_t (aligned to 8 bytes), then the struct will instead have seven bytes of trailing padding so that it reaches 8-byte alignment [2].

To reduce the amount of padding a struct will need, structs should always be laid out with data members in order of descending size. Note that in C++, you can explicitly tell a compiler to not add padding by wrapping a struct in: "#pragma pack(push,1)" and "#pragma pack(pop)".

Practice Problem 2.10: What is the optimal way to order the data members of the following UnstructuredData struct to reduce or prevent unnecessary padding? What will be the runtime size of this data structure after your changes compared to if it remains unchanged?

```
struct UnstructuredData{
    char Data1;
    short Data2;
    int Data3;
    char Data4;
};
```

Solution: Without any changes, UnstructuredData will receive a byte of padding after Data1 so that the Data2 is 2-byte aligned. This will automatically make Data3 4-byte aligned, so no padding is needed there. But we will need three bytes of trailing padding so that if this struct is put into an array then the next entry will be aligned. The resulting size of the struct will be:

1-byte (char) + 1-byte (padding) + 2-byte (short) + 4-byte (int) + 1-byte (char) + 3-byte (padding) = 12 bytes

To reduce padding, we simply order the struct with its data members in decreasing size order. As shown in the StructuredData struct.

```
struct StructuredData{
    int Data3;   // large
    short Data2; // medium
    char Data1;  // small
    char Data4;  // small
};
```

The resulting struct does not need any padding; and the final size is only eight bytes, which represents huge savings when compared to the original size of twelve bytes.

There are many additional ways that we can reduce the size of data types, memory optimization is a major part of gameplay programming. For example, bools use eight bits even though they only hold true/false information which in theory only needs one bit's worth of data. In the earlier bitwise operations section, we explored how a byte can be used as a bitfield of bit flags. The C++ compiler actually provides a way to do this automatically for us (though this behavior is compiler dependent). We can declare bools as bitflags like this:

```
bool myFirstBool : 1;
bool mySecondBool : 1;
```

Each line's suffix ": 1" tells the compiler to try to only use one bit for each bool. If we have a long list of bools then every eight bools will only use eight bits instead of sixty-four. Though it could save a lot of memory, this approach has the following limitations:

- The compiler may disregard your request.
- Multiple bools will share an address, so if you set a data breakpoint on one it could trigger when another is changed.

- Writes to bitfields are NOT atomic (they're a read-modify-write) which could cause issues in multithreaded environments. The problem case would be that if we have several of these bools, two processes could think that they are working on separate bools while actually affecting the same shared bit field.

We'll further discuss concerns of multi-processing in its own chapter, but I wanted to provide some of these considerations early because tradeoffs of this nature are a common theme in gameplay engineering. When you're making a game, there is rarely one correct solution; but rather several options that we can compare to find the best fit for our particular needs.

MEMORY REGIONS

Memory is stored in different regions based on its usage. You should be aware of the following memory regions and their properties.

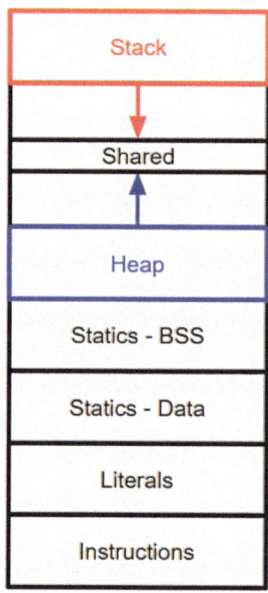

FIGURE 2.6   The stack expands downward; the heap expands upward.

Stack Region:

- Lifetime: temporary, stores locals.
- Size: grows and shrinks based on stack memory allocation and calling nested functions.
- Fastest memory, highest memory addresses.

Shared Region:

- Shared memory is primarily known for its use to concurrently map into the address space of more than one process. This is an uncommon use case in game development because most games are just a single executable (only one process). Though it does have some uses, for example some teams use this region for communication between the retail game launcher and the game itself when passing exception information and logs in the event of a crash.
- In a single program, shared memory can be used to reference the same physical memory from multiple virtual addresses, creating cheap duplicates. This is useful for things like a ring buffer, which can avoid the hassle of handling wraparounds by mapping the buffer twice in a row.

Heap Region:

- Lifetime: managed by program.
- Size: grows and shrinks based on memory allocation.
- Slower access than the stack.

Statics Region:

- Lifetime: duration of program.
- Size: fixed and known at compile-time.
- BSS segment: holds uninitialized global and static values.
- Data segment: holds initialized global and static values.
- To remember the difference between BSS and data I recommend the mnemonic: "BSS is uninitialized and therefore its starting values are bullshit (aka garbage)".

Literals and Instructions Regions:

- Stores compiled code and read-only string literals.
- Typically, the lowest memory addresses.

Practice Problem 2.11: Below are a bunch of values, state the memory region that each will reside in and why.

```
int globalVarA;
const char* globalVarB = "Hello Mrs. Clown, may I have a
bowl of soup?";
  void TestFunction() {
      int localVarC = 42;
      int* localVarD = new int();
}
```

Solution:

- A will be in the .bss section of static memory since it is uninitialized, but it will get a value of zero afterward since globals are zero-initialized (unless they are of type auto).
- B will be in the read-only area of data segment of the statics region.
- C will be in stack memory.
- D itself will be in stack memory, but its dynamically allocated memory will be in heap memory.

CACHING

Caching is a process of optimizing memory usage to minimize how frequently we search main memory. Whenever your program asks the processor to access a section of memory, the processor brings a chunk of memory, called a cache-line (typically 64-bytes), to a temporary storage location called a cache. From the cache, the processor returns the part you want. When your program asks to access data again, it will check if the requested data is within the cache. If so, it can bring you memory from there. If not, this is called a "cache miss", and the processor falls back to a much slower main memory load in order to find your data. It is extremely worthwhile to design our programs to reuse the same sections of memory repeatedly to prevent cache misses.

Practice Problem 2.12: If a program asks the processor to grab the tenth byte of memory, what range of memory addresses will be accessed?

Solution: The processor will grab one cache line's worth of memory starting at the closest cache line interval. On a 64-bit system, the first cache line interval includes byte indices 0-63. The second includes indices 64-127. The tenth byte (index nine) is within this first interval, so the processor will pull those bytes (indices 0-63) into the cache, and then return the desired byte.

Because reusing cached memory can lead to such amazing time savings, most computers are designed with multiple caches. If the desired memory is not found in the fastest "L1" cache, then the processor will check the slower but larger "L2" cache. The PS4 has multiple L1 and L2 caches, the PS5 has an L3 cache, and some computers even have L4 caches. The time difference between accessing data from each cache level is roughly a factor of three: so L1 access is 3x times faster than L2, or 27x times faster than L4. The lower numbered caches may be faster, but they hold less data.

The processor usually maintains a list of flags to track if cache lines within the caches have been changed, to determine if those changes need to be propagated to main memory. A main memory update is typically done when a cache's data is updated, which further extends the time cost of a cache-miss.

In a multi-processor environment, multiple cores can share the same cache; but if they affect its data, then they need to mark it as dirty which can lead to some nasty concurrency problems. To better support multi-processing, platforms often include several caches of the same level (such as the PS4 which has two L2 caches). Likewise, processors can benefit from decoupling instruction storage into a separate cache so you might see those instruction-specific caches referred to as "i-caches", as opposed to "d-caches" which are for data. Some caches, that are designed for both data and instructions, are called unified caches.

In addition to reducing cache misses, applying cache locality understanding to our program design can allow us to use powerful hardware operations called SIMD instructions. SIMD or "single-instruction-multiple-data" instructions allow us to perform multiple computations at once. The exact capabilities of SIMD instructions are going to vary based on the hardware architecture, but great game programmers are never afraid to get down to the metal! The popular Unreal Engine, has many operations that include SIMD instructions including vector maths, matrix operations, and collision detections.

Knowledge of cache locality can also inform our software design to avoid the following pitfalls:

- **Cache thrashing**: when frequent cache misses require frequent loads from main memory.
- **False-sharing**: when multiple cores are editing the same cache line and are frequently setting the dirty flag due to constant cache misses. This happens when a process is grabbing memory for itself and scoops-up, then edits, a cache line that includes memory needed by another process.

Practice Problem 2.13: Imagine you are optimizing a game that manages thousands of EntityData structs (defined below). The game is currently designed with a huge array containing a struct for every entity of which there are thousands. In the game, we frequently need to cycle through that huge array to access entity info to check things like how many are dead or which entities are nearby a position. How could you apply the principles of data locality to this design in order to increase the speed of these operations?

```
struct EntityData {
    int health;
    Vector3 position;
    string otherStringData;
}
```

Solution: This classic problem is often called an Array of Structs (AOS) versus a Struct of Arrays (SOA). Iterating through the structs to look for a particular kind of data will require jumping over lots of irrelevant data, resulting in far more cache misses than if the health floats were arranged beside each other. This problem will become much worse if additional data members are added to EntityData struct.

The solution is to replace the array of structs with a struct of arrays like this:

```
struct EntityData {
    int [] healths;
    Vector3 [] positions;
    string [] otherStringDatas;
}
```

Now we can iterate straight through all of health values, performing the desired operations with far less cache misses. Additionally, if we are performing the same operation on multiple values (such as decreasing all entities' health by five points) then we now have the added affordance of using SIMD instructions because the data is contiguous.

The translation lookaside buffer (TLB) is an additional type of cache that maps addresses to virtual memory pages. When a computer runs out of RAM pages it can create virtual memory pages to simulate a larger memory capacity. The RAM still has the same underlying maximum storage capacity, so when it swaps out real pages with virtual pages it puts the real data into main memory and uses the TLB to remember where it put them. The YouTube video "Virtual Memory: 11 TLB Example" by David Black-Schaffer includes a nice demonstration of the process [3]. Pages on a 64-bit system are typically all the same size of 4KB, but you can alternatively setup a TLB to segment virtual memory into non-uniformly sized allocations called segments; these two approaches are called paging and segmentation respectively.

ALLOCATION

An allocator is an algorithm for partitioning memory for usage. Good allocators have high speed and low fragmentation. Fragmentation refers to inefficiencies in memory usage. There are two kinds of fragmentation:

- **Internal fragmentation**: when you give a client more memory than they need, leading to wasted unused space within the allocations. It's like giving a very skinny man a super long bench all to himself, there is extra room on that bench that is wasted since the occupant is so skinny. Think "wasted space within allocated memory".
- **External fragmentation**: when you have memory available, but you can't allocate it because the free memory is interspersed between used memory, causing it to be too small for the client to use. It's like telling a skinny person to sit in the center seat of a row of three seats. But then a fat, two-seat wide, person requests a seat. Even though you have two free seats (one on either side of the skinny person), they are not contiguous so you can't fit the fat person anywhere. Think "wasted space between used memory".

The main steps of a basic "malloc" allocation are listed below [4]:

1. Determine the size of requested memory (including superclasses and vtables of a class if applicable)
2. Enter a critical section for memory allocation
3. Begin allocation logic (this varies but I will soon explain a few common approaches)
4. If necessary, execute a system call to move the "break" (to grow the heap). Note: this requires a context switch from user mode to kernel mode.
5. If necessary, the kernel may need to write to disk.
6. The memory is returned and, back in user mode, the constructor is executed if applicable.

These steps can be rather expensive, and there are many allocation algorithms designed to optimize the process. I recommend the lecture, "Dogged Determination" by Jason Gregory on YouTube for a primer on some of these designs [5]. Keep in mind that in most allocation algorithms, memory is not cleared once freed, that takes too long. Instead, memory is simply recorded as free and then provided with its garbage data when the client requests memory. Below are some common allocation strategies that you should be aware of. As you read through each approach, consider its benefits and drawbacks with respect to speed and fragmentation.

- The buddy allocator first divides memory into big blocks that are then broken into smaller blocks that will be re-absorbed when they are freed. The sizes of allocation are somewhat arbitrary, so external fragmentation chances are high. But there is less internal fragmentation than a traditional slab allocator since we have more flexibility to give requesters the exact amount of memory that they need.
- The stack aka "LIFO" aka "Ring" aka "Bump" allocator (it has many names) is used for situations that have predictable build-up and tear-down sequences. There is zero internal or external fragmentation because we know exactly what sizes are needed and because the memory deallocation order mirrors the memory allocation order. Sometimes it's designed to grow the stack in two directions.
- The relocatable heap can move allocated blocks after deallocations, but it must manually update any external pointers that were tracking addresses in the allocated blocks. This is useful if we want to extend an allocated object beyond the size, we originally allocated for it (aka reallocate).
- The pooled allocator provides small, usually fixed-size, allocations but does not free them to system memory until the entire pooled allocator is destroyed. Instead, when a single allocation is no longer needed, it is marked as "free" internally within the pool for the pool to reuse in anticipation of a request for the same memory size. This approach is ideal for very many, very small, allocations that are expected to have short lifetimes. Like the stack allocator, it has very low internal fragmentation, but this design allows reuse of allocated memory at the cost of some external fragmentation. Often in strategies like these, a "free list" is used to link which allocations in a pool are currently marked as free.
- The slab allocator is sometimes framed as a specialized pooled allocator where allocations are always of a fixed size. Because the allocations have known sizes, they have known bounds which makes the allocator very fast at swapping slabs in and out. However, because we must round the requested memory amount up to the closest slab size, we are basically guaranteeing internal fragmentation for the benefit of zero external fragmentation. Slab allocators are ideal when all allocations are the same size.
- The frame allocator is a stack allocator that gets wiped each frame.
- The static allocator makes allocations at the start of the program during a defined setup timeframe and then no further allocations are made throughout the duration of the program.

Practice Problem 2.14: You require an allocation algorithm that will allow for several AI to save data that they can refer to during their lifetime. In this context, "AI" refers to an NPC enemy. There is a limit to the number of AI we plan to support but it will vary by AI type. And as AI instances die, they will be replaced by new AI that gradually spawn into the game. There is a set amount of how much data each AI type will need but that amount varies drastically between the different types of AI. Which allocation design would be a good fit for this scenario? Discuss the benefits and drawbacks of your selection with specific reference to both kinds of fragmentation.

> **Solution**: Though there are many viable approaches, I would expect candidates to select a pooled / slab allocation approach for this scenario. A slab allocator can be set up for each of the known AI types and pre-allocated since we know exactly the maximum number of AI that could be simultaneously present for each type. Since each slab would provide us with exactly the right sized allocation for each AI type, we would have no internal fragmentation.
>
> When an AI dies, its allocation will become unused, leading to external fragmentation. However, since each AI of a given type requires the same amount of data, the next AI of that type which spawns can replace the recently deceased AI's allocation.

UNSORTED TOPICS

i++ vs ++i

If the ++ comes before the variable, then the value is incremented before it is accessed. If it comes after, then the value is first accessed and then incremented.

Practice Problem 2.15: State the value of B and C after the following lines of code are executed:

```
C = 10;
B = C++;
```

> **Solution**: The ++ comes <u>after</u> the value is accessed. So, B is first assigned C's value of 10, then C is incremented. The result is B=10 and C=11.

Version Control

Perforce is the industry standard for game development repository management. I would recommend using Perforce for a personal project to gain some familiarity with the software. On an interview you might be asked behavioral questions about your workflow and software development processes. Being able to discuss Perforce and version control principles at that time will be helpful for your candidacy.

Some topics I would advise studying on this matter include:

- How is a perforce stream different from a git branch?
- What is the merge-down and copy-up principle?
- How does file ownership work?
- What is a merge conflict and how can it be resolved?

Practice Problem 2.16: What can a development team do to reduce the frequency and severity of merge conflicts?

Solution: A development team should commit changes frequently, use feature branches, and keep change lists small and focused. Clear communication and code reviews can help avoid overlapping work, while automated conflict detection through continuous integration (CI) systems can catch issues early. Regularly merging feature branches into the main branch and adopting a clear branching strategy is also beneficial. Whenever possible, use non-binary, human-readable data to facilitate easier conflict resolution as a merge rather than choosing one version over another. When you do need to use binary assets, consider locking them to signal to other developers that you've taken temporary ownership of the data.

WORKS CITED

[1] "Zero to One," *Float*, Ridiculous Fish. [Online]. Available: https://ridiculousfish.com/blog/posts/float.html. [Accessed: 11-Aug-2024].

[2] E. Lan, "Why is the compiler adding padding to a struct that's already 4-byte aligned?," *Stack Overflow*, Jul. 2, 2017. [Online]. Available: https://stackoverflow.com/questions/45463701/why-is-the-compiler-adding-padding-to-a-struct-thats-already-4-byte-aligned. [Accessed: Aug. 4, 2024].

[3] D. Black-Schaffer, "Virtual Memory: 11 TLB Example," *YouTube*, Dec. 8, 2020. [Online]. Available: https://www.youtube.com/watch?v=95QpHJX55bM. [Accessed: Aug. 27, 2024].

[4] J. Rodriguez, "Code for Game Developers - Anatomy of a Memory Allocation," *YouTube*, Jan. 1, 2024. [Online]. Available: https://www.youtube.com/watch?v=c0g3S_2QxWM& list=PLW3Zl3wyJwWPmA00yqu9wiCREj4Of_1F8&index=17&ab_channel=Jorge Rodriguez. [Accessed: Aug. 4, 2024].

[5] 'XXI SINFO - Jason Gregory - Dogged Determination," *YouTube*, uploaded by SINFO, Aug. 21, 2024. [Online]. Available: https://www.youtube.com/watch?v=f8XdvIO8JxE& ab_channel=SINFO. [Accessed: Aug. 27, 2024].

C++ Topics

3

This chapter will provide a broad sampling of several C++ topics, focusing on areas that are most targeted in technical interviews. Like the rest of this book, it's only a starting point or reference guide and not an exhaustive education. You are strongly advised to study through Seth Meyer's *Effective C++* series and watch talks from CppCon. I recommend the work of Herb Sutter in particular, he's a great teacher.

THE NEW KEYWORD

New vs Malloc

C++ is often called "C with classes". C++ added a lot of object-oriented thinking to C and the distinction is commonly explored in interviews through asking candidates to compare the "malloc" and "new" keywords. The "malloc" and "new" keywords both allocate memory, and they correspond with the deallocation keywords "free" and "delete", respectively. I recommend the mnemonic "Free Memory" to remember that "free" corresponds with the "m" in "malloc" rather than the "n" in "new".

C-style allocations

- malloc → free

C++ allocations

- new → delete
- new[ ] → delete[ ]

Here are some differences between new and malloc:

new:
- new is an operator.
- new allocates memory and calls a constructor for object initialization.
- The return type of new is an exact data type.

DOI: 10.1201/9781003565550-4

malloc:
- malloc() is a library function.
- malloc allocates memory but then does not call a constructor.
- The return type of malloc() is void*.

If you think of C++ as "C with classes" then it will make sense that the object-oriented functionality, like calling constructors and returning object types, are aspects of the C++ "new" keyword. The "new" keyword is literally the newer keyword that only started to be used once this additional functionality was introduced.

Placement New Operator (In-Place New)

The "new" operator returns the amount of requested data from wherever the heap allocator chooses. However, sometimes we want to control where an object is created. This is the case in a common strategy where we pre-allocate a large amount of data, and then we work directly with the data. C++'s placement-new operator empowers us do this with the following template:

```
type * myPtr = new (ADDRESS) TYPE(INIT)
```

In the template, ADDRESS is the desired destination address, TYPE is the variable type (which implicitly communicates its size), and INIT is the initialization argument list. Below is an example of the template filled in to create a pointer to a new int that is written to x's address and initialized with value 10.

```
int x;
int* myPtr = new (&x) int(10);
```

THE VIRTUAL KEYWORD

The term virtual has different meanings depending on the context. In this section we will explore a few uses that commonly come up in interviews.

Virtual Pages

We already discussed the role of the TLB simulating additional memory capacity using "virtual" pages. See the previous TLB section in the Fundamentals chapter for more information.

Virtual Functions (and vTables)

Functions can be marked as virtual so that subclasses can override their functionality. Virtual functions have additional overhead because they are stored in the vTable. Their benefit is sharing the signature from a class to its subclasses. And, if pure virtual, ensuring that subclasses implement the function (as we will discuss later). Any class that has at least one virtual function or virtual base class is polymorphic. Only those types have a virtual function table aka "vTable" in their data layout. Each virtual class has a vTable which contains a map from all of its virtual functions to their definitions.

Virtual class memory overhead includes:

- One vTable per virtual class.
    - vTables will have an entry for every virtual function (including pure virtual functions) that its class can call. Those entries will each consist of a pointer to the location of the most specific function definition. All the vTables are stored in the static read-only data segment of memory.
    - To find a class' vTable, the linker typically uses the first non-inline, non-pure, virtual function declared in the class, sometimes called the "key" function.
- One vPointer per instance of a virtual class. The vPointer is typically placed in data at the beginning of the object allocation before any member data.

I should mention that the C++ standard does not define how virtual functions are implemented, only how they should behave. So, the exact virtual memory layout may vary by compiler, but the above layout is what we expect to see.

Practice Problem 3.1: Draw the vTable(s) and vPointer(s) that would be constructed at runtime for the following program.

```
class B {
public:
  virtual void Hello(){}
  virtual void Howdy(){}
};
class D : public B {
public:
  virtual void Hello(){}
  virtual void Goodbye(){}
};
int main(int argc, char* argv) {
  D *d1 = new D();
  D *d2 = new D();
  B *b1 = new B();
  return 0;
}
```

Solution: Each of these classes contains at least one virtual function so each will have a vTable. The vTables will have one entry for each of their virtual functions.

```
B's vTable → [ &B:Hello, &B:Howdy ]
D's vTable → [ &D:Hello, &B:Howdy, &D:Goodbye ]
```

We will see a vPointer for each instanced class.

```
d1's vPointer → D's vTable
d2's vPointer → D's vTable
b1's vPointer → B's vTable
```

This question was inspired by the dexterous' question "How many vtable and vpointers will be created in this example?" on Stack Overflow [1].

Dynamic vs Static Dispatch

Static dispatching refers to when a compiler knows exactly which function definition will be used for a function call. Dynamic dispatch refers to the process of determining which function definition a function is referencing at runtime. Virtual function tables are an example of dynamic dispatching. Marking a function as virtual is telling your compiler to dynamically dispatch the function call. We can deliberately perform a static dispatch by specifying the class type before the function invocation such as `ClassName::functionName()`.

Will Virtual Function Calls be Inlined?

Inlining is a compiler optimization wherein the compiler adds the code of a function directly into wherever the function is called. We will further discuss inlining further into this chapter, but it is important to know now that virtual functions will only sometimes be eligible for inlining. Whenever a virtual function is called using a base-class reference or pointer, it cannot be inlined (because the call is resolved at runtime). But whenever a compiler can determine the exact type of an object making a virtual function call it will inline the call.

Virtual Base Classes

When dealing with complex class hierarchies in C++, you may encounter situations where a class inherits from multiple subclasses, each of which inherits from a common base class. This can lead to the duplication of the base class's data members in the derived class. To address this issue, C++ provides a mechanism called virtual inheritance.

Practice Problem 3.2: Imagine you have a class named A which implements a function. Class A has two derived sub-classes, B1 and B2, which both override A's implementation of the function. Additionally, let's imagine that a fourth class exists, called C, which is both a subclass of B1 and B2. When C calls the function, will it use B1's definition of the function or B2's? How can this design be improved?

Solution: This classic interview question is named the "Diamond Inheritance" problem because the class hierarchy can be drawn as a diamond with A at the top, B1 and B2 in the middle, and C at the bottom. The result of this situation is that A's data members are inherited twice into C (once from B1 and once from B2).

C++ requires you to specify in the call to C's function whether the implementation of B1 or B2 should be used. The invocation looks something like this: C.B2::Function() which is a static dispatch indicating that we should call B1's override of the function. This resolves the ambiguity but it does not resolve the data duplication.

To prevent duplicating A's data, you need to define subclass B1 and B2 with virtual inheritance, which looks like this: `class B1 : virtual public A{ }`. This will prevent data member duplication as long as both B1 and B2 are both defined with virtual inheritance. In general, it's good to use virtual inheritance whenever you are using a derived class as a base class.

To recap, here's an example where two subclasses correctly use the virtual keyword to prevent duplication:

```
class SubClass1 : virtual public BaseClass {};
class SubClass2 : virtual public BaseClass {};
class NewClass : public SubClass1, public SubClass2 {};
```

Pure Virtual Functions

Pure virtual functions aka "abstract" functions are simply virtual functions that do not declare an implementation. Pure virtual functions must be implemented by subclasses.

Interviewers may ask you to write out a pure virtual function definition like this example:

```
class Base {
public:
    virtual void FunctionName() = 0;
};
```

A virtual function makes a class abstract and forces its children to implement it otherwise they will also be abstract. The function is not truly "abstract" itself. Though it's rare, pure virtual functions <u>can</u> define bodies so that derived classes can call their base class's implementation when implementing the pure virtual function.

The closest C++ comes to a truly abstract function would be a pure virtual function with no body. This way C++ is a bit more malleable than other languages like C#. In C#, you cannot define a body for an abstract function, so you are forced to either provide a shared definition for subclasses or mandate the function is implemented in subclasses. But C++ allows you to do both of these options: marking a function as pure virtual is the mandate and yet you can still define a body for derived classes to call.

A program will never dynamically dispatch directly to a pure virtual function. This is because if a class has a pure virtual function, then it is an abstract class. Since it is abstract, it will never be instanced. If a subclass of the abstract class exists, then it will have to define an implementation for any abstract functions and those implementations will be dynamically dispatched to instead of the abstract class' implementation (if any). Pure virtual functions are only statically dispatched (invoked manually). A vTable will have an entry for each pure virtual function in its associated class, but they're typically just placeholder, generic fail-state functions, defined by the compiler.

Virtual Destructors

You need to mark a class' destructor as virtual if you will destruct a derived object of that class using a pointer with the type of its base class. Without a virtual destructor, memory can leak. Here's an example:

> If Pet* p = Dog, and I delete p (which has no virtual destructor), it will destruct the Pet data but not propagate the call to Dog's destructor. This can lead to an improper cleanup of resources, potentially causing resource leaks or undefined behavior.

Practice Problem 3.3: When does a class need a virtual destructor?

Solution: This is an extremely common interview question. You should memorize that, "technically", your function does not need a virtual destructor if all the following statements are true:

- Your class is never instantiated on the heap.
- You have no intention of deriving classes from the class.
- You never refer to an instance of a derived class using a pointer of type baseClass.

I emphasized "technically" because, in practice, you should probably early-implement virtual destructors, just in case the class' usage changes over time. The wise Scott Meyers advises that if a class has any virtual function, it should have a virtual destructor unless it's not designed to be a base class or not designed to be used polymorphically [2].

THE STATIC KEYWORD

The static keyword can be used in the context of static functions and static variables, but in both of these cases the keyword means the same thing: one-time allocation in a static storage area within scope for the entire duration of the program.

Static Classes

C++ does not have static classes, but in the languages that do support this feature, a static class is one that cannot be instantiated and can contain only static members. Remember that because its members are static, their data is stored in the static area rather than on the stack.

Static Functions

Static functions within a class are instanced only once, then shared by all instances of the class. Because there is no class instance associated with a static function you can access it without needing a class instance.

Its unique properties are that…

- It can't directly access the non-static members of its class.
- It can't be declared const, volatile, or virtual.
- It doesn't need to be invoked through an object of its class, although for convenience, it may. Usually, static member functions are called using the class name: ClassName::function( ).

Static Variables

Static variables within a class are instanced only once and shared by all instances of the class. Because there is no class instance associated with a static variable you can access it without needing a class instance. These are basically global variables, but there is an organizational benefit to making a variable static so that it can be strongly associated with a class.

Primitive static variables that are not const-initialized and are instead always zero-initialized. In the below example, I have set the names of some variables to contain the substring "zeroinit" to convey that they will be zero initialized and I named one "undef" to convey that it will be undefined.

```
static int zeroInit1;
int zeroInit2;
int main(){
    int undef;
    static int zeroInit3;
}
```

Notice that even though zeroInit2 is not static, it is still zero initialized. This is because it is a global variable. Remember that zero initialization is only for primitive types, there is no automatic zero-initialization for user-defined types.

> **Practice Problem 3.4**: What is the difference between a static member variable and a static local variable in C++?
>
> > **Solution**: A static member variable belongs to a class and is shared between instances of that class. A static local variable is implemented within a function and will retain its value across multiple calls to that function.

AUTO-GENERATED FUNCTIONS

Six functions are auto-generated by compiler, by default they're all public and you need to learn them because several studios require candidates to recite these functions from memory. As a mnemonic, I use the phrase "DC and Marvel comics" to remember the categories of destructor / constructor, move function, and copy function. Each of these three categories has two functions. All six functions are listed below, by category, and accompanied by examples of what they look like in code.

"DC…" – destructor / constructor

- default constructor – Foo::Foo(){}
- destructor – Foo::~Foo(){}

"…and <u>M</u>arvel…" – <u>m</u>ove functions

- move constructor – `Foo(Foo&&);`
- move assignment operator – `Foo& operator=(Foo&&);`

"…<u>C</u>omics" – <u>c</u>opy functions

- copy constructor – `Foo (Foo const&){}`
- copy assignment operator – `Foo & operator=(Foo const&);`

When Are They Not Auto-Generated?

The six functions will not be auto-generated in the following cases:

- When you implemented them explicitly.
- When you prevent them explicitly (such as by using the "delete" keyword).
  - Example: `Foo(Foo const&) = delete;`
- When a compiler can't determine how to generate them, such as if Dog is-a Pet and Pet has no copy constructor. In this case Dog is not going to automatically have a copy constructor because it would not know how to deal with Pet's members during a copy.

Rule of Three

The "rule of three" states that if a class defines any of the following then it should probably explicitly define all three:

- Destructor
- Copy constructor
- Copy assignment operator

If your class is using move semantics, then the move constructor and move assignment operation should be explicitly defined as well. In these cases, the rule is referred to as the "rule of five" to account for the two additional functions.

Assignment Operator

Some considerations for when you are implementing your own assignment operator.

- An assignment operator is only called if the object has already been constructed, otherwise the constructor is called.

- A class' assignment operator needs to free allocated memory that was previously held by the object. For comparison, this is unnecessary in the class' constructor, because, since the object has not yet been constructed, no memory has been reserved. This distinction is the key reason we need two both a constructor and an assignment operator. The only exception is if a class calls the assignment operator on itself, in which case we usually want to identify this unique scenario and do nothing [3].

CASTING

You are expected to memorize the following four C++ cast types. To remember them I use the mnemonic: "coders and designers rarely sleep".

const cast: used to override the const and/or volatile keyword.

- Modifying a previously const var is generally undefined.
- Usually indicative of a design flaw, we shouldn't be casting to and from const.

dynamic cast: used to polymorphically convert pointers and references.

- No cast is needed to convert from subclass to base class. You only need to dynamic cast if converting from a base class to a subclass.
- Only valid if RTTI is enabled for the compiler, which is usually the default.

reinterpret cast: R for "Risky", c-style, no warnings, not type-safe.

- You typically want to avoid using this (except maybe for casting a void* from malloc).

static cast: S for "Standard", like c-style cast but with compile-time checks and more safety.

- Used for most normal type conversions.
- It has no runtime checks, only static, compile-time checks.

C-style cast: the C-Style cast should generally not be used. It tries some of the casts above, ending with the reinterpret_cast. Note that dynamic_cast is never considered when using a C-style cast. Here is the order of C++ casts that a C-Style cast attempts to do:

1. const_cast
2. static_cast
3. reinterpret_cast

Practice Problem 3.5: You have a float value that you need to convert to an int type, which C++ cast should you use?

> **Solution**: Static cast.
>
> ```
> static_cast<int>(floatValue);
> ```

POINTERS

Smart Pointers

These are the three kinds of smart pointers that candidates need to memorize. Sadly, I don't have a mnemonic for these, maybe you could come up with one!

Unique: used when you only need a single reference to an object.

Shared: used when you do not want de-allocation until all references to it are gone.

- Note that because a shared pointer needs to increment its number of current users atomically, it must lock if shared between threads; This could be a performance issue.

Weak: used to point to an object if it's still around, but this pointer is not counted when determining if a resource is in use.

Practice Problem 3.6: When should smart pointers be used instead of C-style "dumb" pointers?

> **Solution**: The answer to most questions of this nature, which pit C++ approaches against C approaches, is going to be "almost always prefer the C++ approach". Smart pointers provide a lot more capability than dumb pointers, including improvements to clarity, exception handling, and ownership. But dumb pointers still have their uses such as in the case of pointer arithmetic.

Should a Function Parameter be a Reference, Value, or Pointer?

This topic is another frequent interview question that asks us to compare approaches. As usual, the answer is: "it depends". Here are some considerations:

> First, we need to consider if we want the function to edit the parameter, which we typically only do to use the parameter as an additional return type. This is sometimes necessary in C++ which does not otherwise support a function having multiple return values. If we need to edit the argument then we can't use pass-by-value, since that will always create a copy, and changes to that value will not propagate to the caller. Therefore, if we need to edit the argument, our only options are to use a reference or pointer. References cannot be null, so they save the function from needing to implement a null check and therefore are generally preferred. If the argument is optional, then a null pointer is sometimes used as a way to communicate that.

An additional benefit of passing arguments by reference or pointer is that we bypass the expensive operation of copying the argument's contents. If the value is smaller than the size of a register (64 bits on a 64-bit system), then it makes more sense to just pass-by-value since any additional overheard required for pointers/references will not provide any benefit.

References vs Pointers

A reference…

- Cannot be reassigned.
- Cannot be null.
- Does not have its own address

A pointer…

- Is its own variable with its own memory and size.
- Supports pointer arithmetic (pointer++).

MISCELLANEOUS C++ KEYWORDS AND VOCABULARY

It's certainly lazy and frowned upon, but some inexperienced interviewers will evaluate candidates by asking them to simply state the definitions of various C++ keywords.

Below, I've listed some of the keywords that come up most often, along with some brief notes on their roles. I encourage you to consult the C++ standard with respect to their exact definitions if any of the terms are unfamiliar to you.

Volatile: "omit from optimization"

Friend: used to declare one class as a friend of another. Classes can access the private and protected members of any of their friend classes. Friendship can also be applied to structs and it works the same way.

- Friendship is not transitive (friends of friends are not friends).
- Friendship is not necessarily bi-directional (just because A is a friend of B, B is not automatically a friend of A).
- Friendship is not inherited (subclasses of friends do not become friends automatically).

Example friend declaration:

```
class ClassName{
private:
    friend class OtherClassName;
};
```

In the above example OtherClassName is declared as a friend of ClassName. You may have noticed that the friend declaration is within the private access specifier but that is irrelevant; access specifiers have no impact on friend declarations.

Const: const is an extremely common keyword that determines if a variable will change its value. Usually, the const keyword precedes a type in an initialization statement. For example:

```
const int x;
```

The above line means x cannot change the value of its integer. However, in the case of pointers, the const keyword can be placed in several different positions, such as in the below example.

```
[A1] int [A2] * [B] myPtr;
```

In this example, the const keyword can be placed in either the A1 or A2 position, and then it can be placed again in the B position. Let's break down what the const value would mean in each of these positions:

- If const is placed at location B, then the <u>pointer</u> is constant. That means we cannot change which memory address the pointer is pointing to, but it does not restrict us from changing the value located at that address. If const is placed on B, then reassigning myPtr's value, such as in the following example, will produce an error:
  - `myPtr = &otherValue;`

- If const is placed at location A1 or A2 then the <u>value</u> is constant. That means we cannot change the value pointed to by the pointer, but it does not restrict us from changing the address that the pointer is pointing to. Const cannot be placed on both A1 and A2 simultaneously, one position must be chosen; most game studios choose either the A1 or A2 position to be the definitive position to use for all of their statements (usually A1). If const is placed on A1 or A2 then reassigning the value myPtr is pointing at, such as in the following example, will produce an error:
  - `*myPtr = newValue;`

Reflection: refers to inspecting and modifying program structures at runtime. It has many use-cases such as compiling code based on non-code data or displaying a property field in a component editor. Unity is a game engine that is well known for its use of reflection to inspect classes for special functions such as "Update" and "Start".

Explicit: forces object creation to use the constructor. This prevents implicit conversions that would otherwise be caused. For example, if a function call provides an int into a parameter that expects a customNumberClass, the explicit keyword can be used to prevent that int from invoking the customNumberClass's int constructor (assuming it has one).

Extern: informs the compiler that a variable or function is declared elsewhere, such as in another file, without defining it in the current scope. This allows for sharing global variables and functions across multiple files. Additionally, extern "C" prevents name mangling by specifying C linkage for functions, ensuring compatibility when linking with C libraries or code.

Include vs Using:

- **Include**: inserts a file's contents into the location of the "include" line. This is used to make your source code file (such as a .cpp) aware of a declaration in another source code file (such as a .h).
- **Using**: tells the compiler that in the next code you are using something, usually a namespace, so you won't have to state the namespace explicitly each time.

Inline: suggests to the compiler that a function should be inlined. Inlining is when a compiler inserts the function's contents directly into the area of code that calls the function. The benefits of inlining are…

- Reducing the overhead associated with setting up and executing a function.
- Preventing a jump to a different area of instructions, which leads to better instruction cache locality.
- Potentially allowing a compiler to perform additional instruction scheduling optimization. It's easier for a compiler to optimize inlined functions, rather than try to optimize the instruction scheduling of two separate functions.

Remember that the inline keyword is just a suggestion to the compiler, and that the compiler may disregard your request. There are "always inline" attributes which, in theory, can force an inline; but even those will be disregarded if the compiler simply cannot inline code (such as in the case of polymorphic or recursive code). Please refer to the previous "Will Virtual Function Calls be Inlined?" section for a discussion on inlining virtual functions.

Union: a user-defined datatype that groups variables so that they can all share one value. This is sometimes useful for shared storage of conditionally necessary variables.

Interface: C++ doesn't have Java-like interfaces, but the closest equivalent would be an abstract class with only pure virtual functions (no bodies) and only static const data members. C++, unlike Java and C#, has multiple class inheritance which allows you to implement multiple interfaces. You can then use `is_base_of()` to assess if an interface is implemented.

C++ BUILD STEPS

You should understand, broadly, how the C++ build process works. Here is a list of each step, in order:

Preprocessing: handles "pound define" directives such as #include, #define, and #if. It processes these directives and produces an expanded source file.

Compilation: translates the preprocessed source code into assembly code or intermediate object code. This involves syntax and semantic analysis in addition to code optimization.

Assembling: converts the assembly code (or intermediate code) into machine code, producing an object file (.obj or .o).

Linking: combines object files and libraries into a single executable or library. It resolves references between different object files, linking function calls and variables to their definitions.

Once an executable has been built, it can be run on target hardware. The "loader" loads the executable into memory and the target hardware's operating system then runs the program.

Practice Problem 3.7: If a function is declared but not defined, which stage of the C++ build process will detect this issue?

Solution: The issue will be detected by the linker which will produce an error indicating that it cannot find the definition for the function when it tries to resolve references to it.

Practice Problem 3.9: If a function is used in code but has not been declared, which stage of the C++ build process will detect this issue?

Solution: The issue will be detected by the compiler which will produce an error indicating that it cannot find the function declaration.

MOVE SEMANTICS

To start, it's important to understand the difference between an lvalue and an rvalue. I recommend the video "lvalues and rvalues in C++" by The Cherno on YouTube as a good primer for the roles of each type [4].

rvalues are...

- Typically used on the right-hand side of an assignment.
- Temporary objects.
- Objects without names.
- Typically non-modifiable.
- Movable, but typically not copyable.

lvalues are...

- Typically used on the left-hand side of an assignment.
- Objects that can live beyond the expression in which they are used.
- Objects that have identifiable locations in memory (referenceable by pointers).

Rules to remember:

- You <u>cannot</u> set an lvalue reference to be an rvalue.
- Not allowed: `int& a = 10;`
- You <u>can</u> set a const lvalue reference to be an rvalue.
- This is allowed: `const int& a = 10;`
- rvalue references have two ampersands instead of one. For example this function takes an rvalue reference: void Foo(int&& bar);

Move semantics were introduced in 2011 as part of C++11 and in some circumstances, they can remove the need of unnecessarily copying data. Instead of copying data (aka copy semantics) we 'move' it (hence the term, "move" semantics). "Move" specifically refers to changing the ownership of an object's data, which means taking its pointers and transferring their ownership rather than the underlying data.

Perfect Forwarding

A function uses perfect forwarding to forward arguments without changing their lvalue or rvalue characteristics. The main idea is that if we have a rvalue as an input parameter, then we are also sending it as an input parameter to the next function we are calling. Hence the name "perfect", implying that we have not changed the argument's characteristics.

REORDERING

Compiler Reordering

The order of reads and writes may be changed by a compiler in pursuit of optimization. This reordering may break "sequential-consistency". To combat this problem, we can use read-acquire and write-release barriers.

Processor Reordering

Same thing as compiler reordering, but now with the processor. To combat this problem, we must use fences:

- **LFence:** ensures prior reads (aka **L**oads) are done.
- **SFence:** ensures prior writes (aka **S**tores) are done.
- **MFence**: ensure prior reads and writes are done.

WORKS CITED

[1] A. Dexterous, "How many vtable and vpointers will be created in this example?," *Stack Overflow*, Mar. 7, 2014. [Online]. Available: https://stackoverflow.com/questions/23170175/how-many-vtable-and-vpointers-will-be-created-in-this-example. [Accessed: Aug. 4, 2024]

[2] S. Meyers, *Effective C++: 55 Specific Ways to Improve Your Programs and Designs*, 3rd ed. Boston, MA: Addison-Wesley, 2005, p. 44.

[3] The Cherno, "std::move and the Move Assignment Operator in C++," *YouTube*, Jul. 19, 2021. [Online]. Available: https://www.youtube.com/watch?v=OWNeCTd7yQE&t=545s&ab_channel=TheCherno. [Accessed: Aug. 4, 2024].

[4] The Cherno, "lvalues and rvalues in C++," *YouTube*, [Online]. Available: https://www.youtube.com/watch?v=fbYknr-HPYE&ab_channel=TheCherno. [Accessed: 5-Aug-2024].

Data Structures

4

This chapter covers some of the most commonly used data structures in gameplay engineering. In this chapter, we'll use Big O notation, such as "O(n)", which is used to express an upper bound of the space, or time, taken up by an algorithm. This is sometimes referred to as the worst-case performance. Similarly, we'll use Big Θ notation to represent a tight bound of the complexity, which can be thought of as the average-case performance.

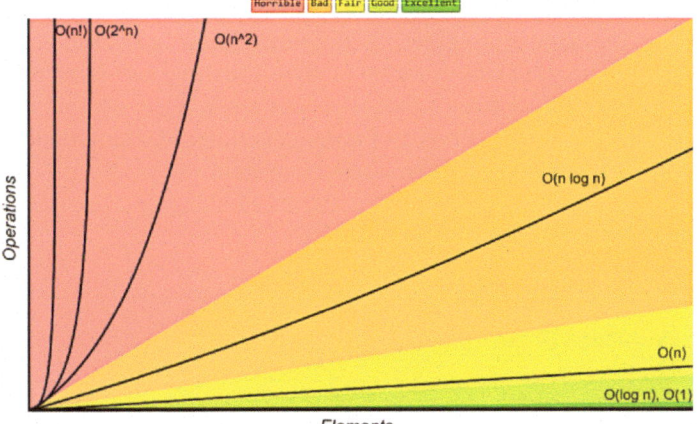

FIGURE 4.1   Diagram illustrating Big O complexities [1].

ARRAY LISTS AND LINKED LISTS

The most common interview questions ask candidates to compare tradeoffs between approaches. Array lists and linked lists can present very different solutions to the same problem, so interviewers love to ask about them! Array lists manage an array internally which means they have lightning-fast, O(1) aka constant time, access speeds. Arrays are the top choice for the vast majority of use cases. And since an array stores its memory

DOI: 10.1201/9781003565550-5

contiguously, arrays have great data locality, meaning that you are less likely to have cache misses. Linked lists, on the other hand, may store the data of adjacent nodes on completely opposite sides of the heap. The key benefit of a list is that you can grow it arbitrarily. On the other hand, if an array extends beyond its capacity, all of its data needs to be reallocated which is an expensive operation.

Terms

- "Array lists" are structures that provide the convenient API of lists (add, remove, etc.) but operate on internal arrays.
- "Linked lists" are "non-contiguous" lists, and since the memory is not together that is why we need to link the entries. This typically means an overhead of at least one pointer per entry.
- When discussed in reference to arrays, linked lists are often just called "lists".

Practice Problem 4.1: What is your favorite data structure?

Solution: The answer is arrays, as they are almost always the best data structure for the job. This question is silly, I just wanted to emphasize a point. The fact that an array's data is contiguous is <u>huge</u>, it's not just a fun fact. The cache locality benefits of arrays are extremely important in real-time programs like games. Learn them, love them.

Practice Problem 4.2: How would you detect if a linked list included a loop?

Solution: This question unfortunately can be a bit of gotcha because it refers to a common algorithm (Floyd's Cycle Detection) which you'll unfortunately just have to memorize. Thankfully it's pretty simple.

- Take two pointers, Fast and Slow, and walk them through the list.
- Slow moves one element at a time and Fast moves two elements at a time.
- If there's a loop, then Fast and Slow will eventually land on the same element.
- If there's no loop, then Fast will reach the end without colliding with Slow.

Time Complexity

Both of these data structures take up the same amount of space, but they have tradeoffs with respect to their speed of various operations. For the following time complexities, we assume that the order of the data does not matter and that the "List" is a singly-linked list. Before you read the speed for each, quiz yourself to see if you can determine the correct speed. Once you read the speed, quiz yourself to see if you can figure out why each is the listed speed.

- Access – Array: $O(1)$, List $O(n)$
    - If you know the index of the array item you need, you can simply index right into the data. However, with a list, you need to traverse the list to reach the element you want. Arrays win here by a huge margin.
- Search – Array: $O(n)$, List $O(n)$
    - Now the array pays the cost of needing to traverse every element, tying both data structures in the race to search.
- Insertion – Array: $O(n)$, List $O(1)$
    - Array: Here is where the big O notation fails to paint a full picture. With arrays, as long as you know that your count won't exceed your capacity, you can plan to have $O(1)$ insertion. You only need to pay the expensive relocation cost if you exceed an array's capacity. Often in game development, we limit the amount of data we are storing. For example, we might limit a map to having at most twenty AI so that we can pre-allocate that much space in an array and know that we won't ever need to expand the capacity.
    - List: With an unordered list, insertion is trivial, lists maintain a pointer to their first element which we can simply point at a new element and have that new element point to the previously first element.
- Deletion – Array: $O(1)$, List $O(n)$
    - Array: An array achieves constant time deletion by placing the data of its highest index into the index to delete, then decreasing its count. This method is called "fast delete." Another approach, "slow delete," identifies the index to delete and then shifts every element above that index down by one. The slow delete is only used when order must be maintained.
    - List: Singly-linked list entries only hold a pointer to the next entry so you need to traverse the list to first get to the entry you want to delete before you can delete it and fix up the pointers, hence $O(n)$. But a doubly-linked list entry has pointers in both directions so, assuming you have the element already, the deletion can be done in constant time.

Practice Problem 4.3: The stated time complexities address scenarios in which the order of the data does not matter. If you were storing numbers that needed to be stored in ascending order, how would that affect the time complexity of each operation?

Solution:

- Access: Speeds would remain unchanged.
- Search: Arrays can now use binary search which would cut the time down to O(log n). But lists would still require you to manually traverse them, so they stay at O(n).
- Insertion: The array remains at O(n) because it needs to shift potentially all of its elements to maintain their order. And if its count exceeds its capacity, we still need to do an O(n) reallocation. The list will now be O(n) because it needs to traverse to the right position.
- Deletion: To maintain order, the array must use the aforementioned "slow delete" which is O(n). And while a deletion maintains ordering for the list, it has to do traversal just like in the unordered case which keeps the speed at O(n).

Ring Buffers

A ring buffer is a data structure that uses an array but, instead of growing when its size reaches its capacity, it overwrites its data, like a snake eating its tail. This structure is ideal for when we want to keep track of only the last few elements and don't mind the older entries getting overwritten. Use cases include logging and streaming.

Queues

A queue is a fundamental data structure that follows the First-In-First-Out (FIFO) principle. This constraint is what allows a queue to provide constant time insertion and removal. Queues can be implemented using either arrays or linked lists. When implemented with arrays, queues often use a ring buffer, which efficiently reuses space previously occupied by elements that have been dequeued. This circular structure allows the queue to maintain a fixed size while still utilizing all available space. If a queue does fill up, its internal array can be reallocated to a larger size. Another strategy is when you're about to enqueue the last element to fill the array, instead enqueue a pointer to a new array as if it were an element, effectively expanding the queue's capacity.

Double-Ended Queues (Deques)

A double-ended queue, or deque (pronounced "deck"), extends the functionality of a standard queue by allowing elements to be added or removed from both ends. Deques are particularly useful in situations where we want to store a fixed number of elements and may need to remove items from one end while adding new ones to the other. For instance, deques allow privileged clients who have already "waited in line" to jump to the front.

It's important to note that while deques offer fast insertions and deletions at both ends, they do not provide the same flexibility as a doubly linked list, which allows for insertions and deletions in the middle of the sequence. This distinction makes deques ideal for use cases focused on first element and final element operations rather than random access modifications.

HASHING & HASH MAPS

Hashing is the deterministic process of converting data into a fixed size value called a hash. In this case, deterministic means that the same data will always convert to the same hash. The main idea behind hashing is that it provides a fast one-way conversion from a potentially very large data type to a hopefully much smaller hash string that can be used for reference. Hashing is not serialization. Serialization is bi-directional, meaning that you can go back and forth serializing and deserializing data. However, you cannot similarly "undo" a hash to recreate input data with only a hash string.

Hashing has an important role in data storage. Hash tables are structures that hash keys into indices, for example C++'s std::unordered_set uses a hash table to provide faster lookup for its array of elements. Hashing speeds up the search process for an array because it allows us to jump all the way to a key's hashed index rather than searching every index for the desired key. Hash maps are an extension of hashtables, which store key-value-pairs (though some languages use the terms interchangeably). The key is hashed and then both the key and value are placed into the structure as an element.

We looked earlier at how a singly linked list had a slow access time of O(n), which required visiting every element. The array was much faster, but it required us to know the index of the data we wanted. Hashing provides a middle-ground tradeoff because the hashing algorithm should provide us with a constant-time method to determine the hashed index of the desired key and then access the element with array-like efficiency.

We prefer hashing algorithms that do a good job of spreading out the indices they generate, because if the algorithm hashed all keys to the same index, the map would have to still search through every key. Therefore, in the worst case, maps have an access time of O(n). But on average, assuming our hashed indices are well spread, we should have a constant time access speed of $\Theta(1)$.

Hashing Functions

As you're probably starting to see, the hashing algorithm can drastically impact a hash-map's performance. One of the key implementation details of a hashing algorithm is how it mitigates and manages collisions which occur if two separate keys map to the same index.

One small way to mitigate collisions is to simply implement the map with an underlying array that has a prime number capacity. At the end of a typical hashing algorithm, you modulo the destination array's capacity with the hash function result:

index = hash % capacity

Prime numbers will lead to better stratified results since they are more likely to have remainders resulting from the modulo.

At some point the array will become "nearly full", which means that its count has reached an arbitrary load factor. Typically, this load factor is 75%, meaning that once the underlying array of the hashmap reaches a count of 75% of its capacity, it is considered nearly full. At this point, for the hashmap to remain efficient, it undergoes a rehash which requires moving all of its elements into a larger capacity structure. Rehashing is expensive, but it can help mitigate additional collisions.

To handle collisions during insertion, there are two main approaches you should be aware of: open-addressing and chaining.

Open Addressing: This method uses a large array of keys for its underlying data structure. When a hash results in an index that is already occupied, we can traverse forward through subsequent indices until we find an available index.

To determine how fast we traverse indices, there are various approaches:

- **Linear probing**: Traverses one subsequent index, one at a time: $1\rightarrow2\rightarrow3$.
- **Quadratic probing**: Jump indices, with increasing jump distance: $1^2 \rightarrow 2^2 \rightarrow 3^2$.
- **Double hashing**: we use a second hashing algorithm to determine our step size.

Chaining: This method uses a small array of linked list pointers as its underlying data structure. Each linked list includes all of the keys mapped to its index. When you hash a key to an index, you find the corresponding linked list and add the key to the list. This approach is easier to grow so it has better rehash prevention, but it faces the typical downsides of a linked list, namely poor cache locality.

Ordered Maps vs Unordered Maps

C++'s std provides both an ordered map (std::map) and an unordered map (std::unordered_map). Ordered maps use balanced binary search trees (like Red-Black trees), while unordered maps use the simpler hash tables we've been discussing.

Generally, with any ordered data structure, you are trading a slower insertion speed for a faster lookup speed. If your game will only store elements while loading a level and then constantly access the elements while it is played, that might be a better fit for an ordered map. Ordered maps are great for narrow and known key domains, while unordered maps are better at handling unknowns.

Another important factor is reference stability, which is a feature of ordered maps but not unordered maps. Reference stability means that your references (and iterators) of an ordered map's elements will stay valid while you insert, add, and remove items. This can save valuable time.

The following time-complexity summary for each structure tells a clear story that while unordered maps perform better on average, they are worse in the worst-case scenario.

OPERATION	ORDERED MAP	UNORDERED MAP
Insertion	$\Theta(\log n) / O(\log n)$	$\Theta(1) / O(n)$
Deletion	$\Theta(\log n) / O(\log n)$	$\Theta(1) / O(n)$
Search	$\Theta(\log n) / O(\log n)$	$\Theta(1) / O(n)$

Practice Problem 4.4: You're developing a system to maintain a real-time leaderboard for a game where you need to frequently add new players and update their scores in addition to rarely retrieving the top players in descending order. Assume that you will implement this system using a map. Should you use an ordered map or unordered map?

Solution: Adding players and updating scores will, on average, be faster for an unordered map. However, retrieving players in-order will be slower for an unordered map since it will need to manually sort its contents each time. Since the ordered retrieval will only happen rarely, the unordered map will probably be a better choice.

TREES

There are many types of trees, but we will only cover some of the most common for game development. Trees are mainly used to implement directed acyclic graphs in order to link nodes that have a one-way relationship or to store sorted data so that they can complete searches in logarithmic time (very fast). Trees are sometimes implemented using an array as an underlying data structure (for cache locality), but they are usually implemented as a sort of linked list where each node is stored separately.

Binary Trees, Quad Trees, and N-ary Trees

A binary tree is a tree where each node has two or less children. Similarly, a quad tree is a tree where each node has four or less children. Quad trees are used to subdivide maps into quadrants, that can then be further subdivided into sub-quadrants, and so on. Lastly, n-ary trees have nodes with n or less children. N-ary trees are commonly used for scene graphs and behavior trees.

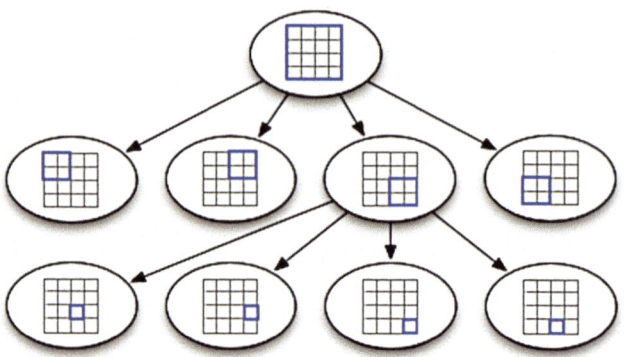

FIGURE 4.2   Diagram of a quad tree [2].

Binary Search Trees

Binary search trees (BST's) are a subset of binary trees that sort their contents so that, for any node: its left child is of lesser value, and its right child is of greater value. As long as a BST is well balanced, it can provide O(log n) search speeds, at the cost of slower insertions and a small amount of additional memory. While there are many balancing algorithms for BSTs, for a coding interview I think the main two you should be aware of AVL trees versus Red–Black (RB) trees.

AVL vs RB

Overall, AVL has faster lookups than RB because it's more strictly balanced. As usual, that faster lookup speed comes as a tradeoff for a slower insertion speed. AVL stores an int of data per node, whereas RB only stores a single bit per node. Both of these trees have a "balance" and "rebalance" algorithm for checking balance and restoring balance respectively. We won't cover the implementation of each algorithm here, because videos on the internet can provide a better demonstration of how they work, but here are the key bullet points to keep in mind:

- Balance speed is O(log n) for both AVL and RB, though RB is $\Theta(1)$ on average.
- Rebalance speed is O(log n) for AVL, yet O(1) for RB.
- AVL stores an int per node (height or balance factor), RB stores 1 bit (red or black)
- AVL is best for lookup (databases), RB is best for insert / delete (maps and sets)

Heaps

Heaps come in two flavors, max heap and min heap, and they're designed to provide fast access to the max or min value of a data set respectively. Like BST's, heaps are another important type of binary tree.

Just like the BST's heaps also have rules that require an inequality between a node and its children. In a max heap, each node is greater than or equal to its children. In a min heap, each node is lesser than or equal to its children. Unlike BST's, all levels of a heap are completely full except for the bottom most level.

Using a min heap for example, it can access its min value in constant-time, O(1). This very fast speed is only possible because the whole data structure is designed around optimizing for this operation. To extract the min value is an O(log n) operation because though the min value can be removed quickly, setting up a new min to satisfy the heap's requirements takes time. The operation to restore a heap's requirements is called "heapify" and basically consists of swapping the new root with its children until it has reached an appropriate new position.

TREE TRAVERSAL

Depth-First Search and Breadth-First Search

Depth-First Search (DFS) and Breadth-First Search (BFS) are two algorithms that interviewers will often ask candidates to recite from memory. These approaches can be used for trees with any number of nodes and for trees with cycles. I have included their abbreviated algorithms below to help you memorize them.

Depth First Search (using stack):	Breadth First Search (using queue):
add root to stack while stack not empty: → pop and discover first element in stack → push undiscovered neighbors of first element	add root to queue while queue not empty: → dequeue and discover first element in queue → enqueue all undiscovered neighbors of first element

You may have noticed that both algorithms look very similar, the only real difference is that BFS uses a queue and DFS uses a stack. Because DFS uses a stack, it can also be implemented using recursion where the system stack is the implicit stack. Below is an example of a recursive solution.

Depth First Search (using recursion):

```
call DFS on root
for each neighbor:
→ if neighbor undiscovered:
→ → call DFS on neighbor
```

Recursive solutions can be faster to write but they are almost never preferred because the additional cost of calling multiple functions can become very expensive.

Since BFS spends more time around the starting node, it's generally a better fit for shortest path algorithms, whereas DFS is a better fit for deep exploration, cycle detection, and backtracking. BFS stores all the nodes at the current level, while DFS only stores the nodes for a particular path, so BFS can have a much higher space cost.

Pre-Order, In-Order, and Post-Order

There are several ways to traverse a tree. The main orders to know are pre-order, in-order, and post-order. These distinctions refer to the order that the root nodes are visited with respect to their children. For example, if the root comes before both of its children, we call it "pre"-order.

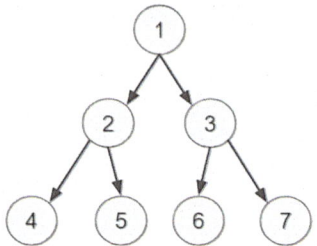

FIGURE 4.3 Diagram of a binary tree.

Nodes from the tree of Figure 4.3 would have the following traversal orders:

- Pre-Order: 1 - 2 - 4 - 5 - 3 - 6 - 7
- In-Order: 4 - 2 - 5 - 1 - 6 - 3 - 7
- Post-Order: 4 - 5 - 2 - 6 - 7 - 3 - 1

These different orders are easy to implement in a recursive DFS where our choice of when to visit the current determines the order. In the below pseudo code, I've implemented a recursive DFS algorithm, it's just missing code to "visit" the current root node. That code can be added to position A, B, or C to achieve pre-order, in-order, or post-order traversal respectively.

```
void DFS(Node current, TreeTraversalOrder order){
    if (current == null) return;
    if (order == TreeTraversalOrder.PreOrder) current.Visit();
    if (current.left != null) DFS(current.left, order);
    if (order == TreeTraversalOrder.InOrder) current.Visit();
    if (current.right != null) DFS(current.right, order);
    if (order == TreeTraversalOrder.PostOrder) current.Visit();
}
```

Practice Problem 4.5: How would you visit the nodes of a tree in level-order? Level order for the tree above would be 1 - 2 - 3 - 4 - 5 - 6 - 7.

Solution: I recommend trying to actually code this algorithm online. It's a good test of your familiarity with tree traversal, which is why interviewers frequently ask it. The website Leet Code has a neat online coding interface where you can run your code and see if it works. Leet Code includes this problem under the title "102. Binary Tree Level Order Traversal" [3]. Just remember to code in C++, because that's the language most gameplay engineers will be tested in.

The basic algorithm is to do a BFS traversal where, for each node, you go through its children from left to right while enqueuing its grand children to handle later.

PATHFINDING

Pathfinding is the process of connecting two nodes in a graph usually by the shortest or most efficient path. While BFS is great for finding the shortest path of an unweighted graph, we need to explore other approaches to deal with weighted graphs. The two main weighted graph pathfinding algorithms you should be familiar with are Dijkstra and A*. Both of these algorithms are designed for graphs that only have positive weights (for negative weights you can look into alternatives like Bellman-Ford). There are great videos on YouTube (which I will link below) that demonstrate these algorithms in action. Below, I have summarized the high-level details you should know for each algorithm.

Dijkstra

The goal of Dijkstra's algorithm is to determine the most efficient path from a particular start node to every other node in the graph. If we have a goal node in mind, we can early-exit from the algorithm once the path to the goal node has been determined. But in those cases that we have a specific goal, A* is generally a better option. Dijkstra really excels when considering multiple goals. It struggles with dense graphs because it evaluates every edge which can be very time consuming.

Helper Structures

- A distance array that tracks the current distance for each node.
- A set that tracks which nodes are unvisited.
- A priority queue (usually implemented with a min-heap) of nodes to quickly access the node of least distance.

Algorithm

1. Mark every node as unvisited by adding it to a set.
2. Mark a starting distance of "infinity" for each node. This is done using both the distance array and the priority queue. Each distance value represents how far the node is from the start node. During the algorithm, we will decrease these values as we find more efficient paths. The start node is special and can be given a starting distance of zero.
3. If node, M, is connected to the start node, replace **dist(node)** with **edge(Start, M)**
   a. In other words, if a node is adjacent to the start node, we replace its distance value with the cost of the edge connecting it to the start node. The result of this first step is that the start node will still have distance zero, nodes adjacent to it will now have distances based on their edges, and remaining nodes will still have infinite distance values.
4. While there remains unvisited nodes, visit the one of the least distance, L, and see if it can relax each one of its neighbors, N, by replacing **dist(N)** with **dist(L) + edge(L, N)**
   a. In other words, we take the node that is closest to our start node that has not yet been visited and we set the distance values of its neighbors. For example, if the closest unvisited node is M and it has neighbor N, then we set the distance of N to be the distance to M plus the distance from M to N. This process is called "relaxing".
5. Once we have visited the goal node, we are done.

You can view this algorithm in the excellent video, "Dijkstra's algorithm in 3 minutes" by Michael Sambo on YouTube [4].

Space/Time Complexity

Where V is the number of vertices (aka nodes), and E is the number of edges, the run-time complexity of Dijkstra is $O(V \log V + E \log V)$.

- The first summand "V log V" comes from doing a log(V) "extract-min" operation on the min heap for a total of V times (once per vertex).
- The second summand "E log V" comes from doing a log(V) "decrease-key" operation to decrease or "relax" the distance of a node in the priority queue for a total of V times (once per vertex).

I should note that during the decrease-key operation we also update the distance in the array, but that is constant time so it's fast enough to not be written down.

The space complexity of Dijkstra is the sum of the priority queue, set, and array which are all O(V). If you include storage of the graph itself, then you must also count the edges which would instead be O(V+E).

A*

A*, pronounced "a star", is designed for finding the shortest path to a specific goal node, rather than to all nodes. Instead of using the true distance like Dijkstra, A* uses a heuristic that estimates the cost from a node to the goal node, this takes up extra storage but can save time. A* is often framed as an extension of Dijkstra because it follows a similar design, the main difference being A*'s directionality and use of a heuristic.

Heuristic

A node's value or "f-cost" is expressed as $f(n) = g(n) + h(n)$, meaning that it is calculated by taking a particular node, n, and forming a sum of that node's g and h costs.

- "g-cost" aka, g(n), is the known cost of the path from n to the <u>start</u> node.
- "h-host" aka, h(n), is the heuristic cost of the shortest path from n to the <u>goal</u> node.
  - To optimize A*, this heuristic should be optimistic.

Helper Structures

- Array(s) to track the current costs for each node.
- The "closed set" which contains nodes that have been evaluated. Usually implemented as a set or list.
- The "open set" which contains nodes that have not yet been evaluated. Often implemented as a priority queue so that you have quick access to the node with the lowest f cost.

- A predecessor array to reconstruct the shortest path once the goal is reached. Each entry corresponds with a node in the graph and indicates the "predecessor" of that node. A node's predecessor is the neighboring node that we took to reach it. For example, if I ascended a flight of stairs (like a normal person): the nth step's predecessor is the step I stepped on just before it, which would be step n−1.

Algorithm

- Initialization: Initialize the start node's g value to zero and h value based on the heuristic. Add the start node to the open set. Every other node implicitly starts as "unvisited" because they are in neither set.
- Loop: While the open set is not empty, select and remove the open set node with the lowest f-cost, with ties broken by the node that has the lowest h-cost. Take the selected node into the iterative step for processing.
- Iterative step: For each neighbor of the selected node:
  a. If the neighbor is in the closed set, skip it.
  b. Update the neighbor's g-cost.
  c. If the neighbor is not in the open set or the new g-cost is lower than the previously recorded cost, update the neighbor's g, h, and f-costs. Set the current node as the predecessor of the neighbor.
  d. If the neighbor is not in the open set, add it.
- Termination: If the goal node is reached, we are done, use the predecessor array to reconstruct the path. If the open set is empty and the goal node has not been reached, there is no path.

Space/Time Complexity

At each step of A* we branch to consider different neighbors. The average branch cost is considered the algorithm's branch factor, abbreviated as b. With each step, we increase the depth of our search. The total depth of the algorithm is abbreviated as d.

Time Complexity

$O(b^d)$. This expression is pretty straightforward, basically we are branching repeatedly (b * b * b * b...) and the total number of branches is d, the depth of our search. A good heuristic will reduce d to provide a faster result.

Space Complexity

$O(b^d)$, we need to store each neighbor at each step of the algorithm, so this is the same as our time complexity: the branch factor multiplied for every step of our total depth. Just like Dijkstra, A* has several variations to alleviate this drawback.

Practice Problem 4.6: In a game with a dynamic environment where edge weights can change frequently, you need to find the shortest path from a starting node to a specific goal node. Which algorithm would be better to use here, Dijkstra or A*?

Solution: Dijkstra will struggle with dynamic edge weights but it's still more viable than A* which really doesn't work without a reliable heuristic. There are some modifications that can be made to Dijkstra to improve performance with dynamic edge weights but even the vanilla version is a better approach here than A*. It's true that in this problem we are looking for a specific goal which is generally a clue to use A*. But in the situation described, A* cannot provide a guarantee to find the shortest path whereas Dijkstra can (even if it might be slower). If we can constrain the scenario to achieve an admissible heuristic (meaning that it always provides an underestimation), despite the dynamic weights, then A* may be viable.

SORTING

While game development uses many sorting algorithms, the main ones I would focus on for interview prep are Quicksort and Insertion Sort. We won't cover the details of their algorithms here, but we will review a summary of their relative tradeoffs.

Algorithm	Time Complexity			Space Complexity
	Best	Average	Worst	Worst
Quicksort	$\Omega(n \log(n))$	$\Theta(n \log(n))$	$O(n^2)$	$O(\log(n))$
Insertion Sort	$\Omega(n)$	$\Theta(n^2)$	$O(n^2)$	$O(1)$

FIGURE 4.4 Space/Time complexity of Quicksort vs Insertion Sort [5].

Quicksort, true to its name, is typically regarded as the fastest sorting algorithm but it does require a non-constant amount of additional space so Insertion Sort may be preferred since that algorithm is more space efficient. Insertion Sort is better for smaller data sets and for nearly sorted data sets. Some approaches use a blend of both methodologies, where Insertion Sort is used to sort the sub arrays that result during the Quicksort process, to achieve benefits from each.

MISCELLANEOUS DATA STRUCTURES AND THEIR PATTERNS

There is an endless amount of data structures to explore within game development. But a few stand out to me as specifically relevant to gameplay engineering. Their relevance will vary based on the exact details of the role you're applying for, but I recommend having some familiarity with each of these. Many of these data structures are coupled with implementation patterns, we'll cover additional patterns in a future dedicated chapter.

Finite State Machines

- Extremely common for tracking animation states such as a state for "Idle" with transitions to "Run" and "Walk" states.
- Sometimes we use layered FSMs. For example, an FSM for upper body animations that operates on top of an FSM for full body animations. This allows us to play separate yet simultaneous animations on the hands and body, such as reloading a pistol while walking.
- Passthrough states (sometimes called conduits) don't allow you to end in that state. If you fail a condition to exit a passthrough, you will fall back to the last non-passthrough state you visited. This approach can drastically simplify the graph structure of complex FSM behavior.
- It's much easier to debug transition logic in FSMs compared to Behavior Trees, because FSMs have explicit transitions between states.
- It's a little harder to visualize more complex if-else statements in FSMs compared to Behavior Trees.

Behavior Trees

- Much more modular than FSMs, nodes are lightweight and only need to return success, failure, or in-progress.
- Useful to manage rapidly changing goals or actions, rather than states like in an FSM.
- These typically scale better than FSMs as it's easier for them to implement subroutines (nested behavior trees).

Blackboards

- Centralized storage for information that can be shared between multiple AI agents.
- Allows AI to coordinate complicated behavior and caches common data to prevent agents from recomputing it too often.

Blend Shapes

- Store common poses for facial features such as a smile or frown and allow animators to tune a dynamic pose as a ratio between baked poses.
- Vastly reduce animation data.
- Vastly reduces the amount of time it takes to animate (animators can just adjust the sliders).

Nav Meshes

- Allow designers to fine-tune which areas of a level are traversable
- Can be dynamically updated (if for example a door opens). This is usually done by toggling a mesh link rather than a runtime re-bake.

Heat Maps

- Allow designers to visualize 2D influences such as sound.
- Can be used to collect useful player data such as marking which parts of a map players are experiencing bugs.

Strings

Strings are a collection of characters that are often used to pass around text. They are very common in interviews. Make sure you feel comfortable manipulating strings in the language you will be using for the interview. And be of any string manipulation features that are provided by that language. Below are a few algorithms I suggest you memorize due to their frequency in interviews. The website LeetCode is a great resource to practice working with strings.

- **Reverse String (Iteratively)**: Use two pointers. One will iterate from the start of the string and move forward until it is half-way through. The other will start at the end and will go backward until it is half-way through. As you go, swap the characters pointed to by each pointer. For example, the first step will swap the first and last character. The last step will either swap the middle two characters, or if the string is an odd number of characters, it will do nothing. Be mindful not to swap the null terminator.
- **Palindrome Detection**: Start two pointers at the middle index and iterate out, checking if equal every step.
- **Palindrome Substring Detection**: Start two pointers at every index and iterate out, checking if equal every step.

MISCELLANEOUS DATA ALGORITHMS

As a final note to this chapter, I want to mention that, while not super common on game development interviews, you should probably study memoization techniques as used in "dynamic" programming problems. The basic approach of these problems is "recursion plus book-keeping". They can be a bit tricky at first and are really best learned through watching worked solutions. The YouTuber NeetCode has an excellent series on this topic, going through many LeetCode problems that use dynamic programming that I suggest you check out [6].

WORKS CITED

[1] "Big-O Complexity Chart," *Big-O Cheat Sheet*. [Online]. Available: https://www. bigocheatsheet.com/. [Accessed: Aug. 11, 2024].

[2] *Quad Tree* [Online]. Available: https://ppujari.medium.com/quad-tree-and-spatial-data-indexing-88b12a87dfd9. [Accessed: Aug. 11, 2024].

[3] LeetCode, "Binary Tree Level Order Traversal," *LeetCode*, [Online]. Available: https://leetcode.com/problems/binary-tree-level-order-traversal/. [Accessed: Aug. 31, 2024].

[4] M. Sambo, "Dijkstra's Algorithm in 3 Minutes," *YouTube*, Aug 6, 2024. [Online]. Available: https://youtu.be/_lHSawdgXpI?si=CgMrX_7lKcZkxsgp. [Accessed: Aug. 11, 2024].

[5] "Array Sorting Algorithms," *Big-O Cheat Sheet*. [Online]. Available: https://www. bigocheatsheet.com/. [Accessed: Aug. 11, 2024].

[6] NeetCode, Dynamic Programming, *YouTube playlist*, [Online]. Available: https://www. youtube.com/playlist?list=PLot-Xpze53lcvx_tjrr_m2lgD2NsRHINO. [Accessed: Aug. 11, 2024].

Patterns

5

PATTERN PITFALLS

Patterns are common approaches to solve common problems. Amongst many programmers, patterns have a bad reputation, and rightly so. They are often misunderstood as a goal rather than just a specific way to fix a problem. For example, I have personally been in an interview where an interviewer asked, "Can you recite the SOLID design principles?" which is a terrible, backwards, question to ask a candidate. Likewise, I have seen this issue come from candidates. For example, I have seen a portfolio where a candidate said, "For this project, my goal was to use the observer pattern". If we are overly focused on a hammer, screws will begin to look like nails. Instead, we need to maintain focus on the actual problem we are trying to solve.

I include this chapter because it is useful to learn about ways in which problems have been solved. But you must understand that these approaches are absolutely not prescriptions and are never a goal in themselves. I also include this chapter because the truth is, this book is ultimately about helping you land a job, and sometimes you need to know the right answers to the wrong questions. Please approach this chapter as you might approach learning the "urban dictionary" of a foreign language. It might be useful to know curse words in another language so that you can identify them, but you probably wouldn't want to say them! Similarly, some of these patterns are not good. But they're useful to know in case we need to discuss them, or in case your interviewer is misguided.

OBJECT-ORIENTED PROGRAMMING

Object-Oriented Programming (OOP) is a programming approach that focuses software design around "meta" objects that contain and operate upon the data a program needs to use. Generally, it's an approach that programmers should limit or avoid, because objects have state which reduces the determinism of a program. When we're writing software, we want to maximize determinism which can be thought of as minimizing state. This

DOI: 10.1201/9781003565550-6

improves a program's predictability which reduces errors and simplifies the debugging process when errors arise. For example, if we can implement a function as member function or a static function, we will always choose the static approach because it communicates that we are operating directly on the data, which is a stricter commitment. As much as possible, we as programmers, want to outsource the ambiguity of a program into data so that we can retain focused, deterministic, and procedural code.

The major pitfall to be wary of with OOP is becoming overly focused on the object abstractions. Like most other patterns, OOP is only useful to the extent it is actually solving a real problem.

Another concern with OOP approaches is trying to model the code in a way that reflects the real world rather than the actual game. For example, a game might have classes Dog and Car. If we were to then add a Horse, that provides transportation, a real-world-model perspective might be inclined to make Dog derive from some sort of abstract Animal class. And then refactor Horse to also derive from Animal. However, if the Horse class is only used for transportation, and it doesn't do things that you need Dog to do, then it may not make sense to structure the classes this way. It may make more sense to simply implement the Horse as a Car, even if that seems a little ridiculous from a real-world-model perspective. The point is to focus on what we are actually trying to (the verbs) rather than focus on how we can abstract the problem into groups of meta objects (the nouns).

The prior horse example illustrates another concern which is that OOP designs tend to overuse inheritance which can lead to monolithic object-based classes. Most game engines are already migrating away from this design, in favor of objects that function more like containers for components of data. For example, you may be familiar with entity component systems (ECS) which strive to isolate data components from the logic systems that use them.

Data-Oriented Programming (DOP) is another architectural paradigm, and alternative to OOP, focused on software design that efficiently organizes and processes data. Mike Acton presented the seminal "Data-Oriented Design and C++" talk at CppCon in 2014, which you can view on YouTube [1]. The talk is a great resource to start learning about DOP approaches. The ultimate message of Acton's talk is that when we are writing software, it's for real hardware and for real problems, and that abstractions can distract us from those real factors and limit performance. He goes through a few examples where object-oriented thinking can lead us to an idealized case rather than the most common case, which should be our main focus.

Resource Acquisition is Initialization

Resource Acquisition is Initialization (RAII) is the idea that acquiring a resource should be tightly coupled with the process of initializing an object that manages the resource's lifetime. For example, the smart pointer's constructor and destructor correspond to the bounds of its memory lifetime. Because the runtime lifetime is determining the runtime memory usage, the memory footprint of program is less determinant. The language Rust is all RAII with super strong types that can be used to prevent leaks. Languages like Jai and Zigg provide RAII alternatives, that favor explicit and manual control of data. The language C++ falls somewhere in the middle, you have RAII objects like smart pointers, but you can also manage data directly through keywords like "new" and "delete".

SOLID Patterns

SOLID patterns are a list of object-oriented coding "principles" that generally favor code flexibility and maintainability over speed and space performance. This is generally not a preferable trade in game development where performance is often the main concern. The main issue with SOLID is that patterns are rarely principles. There's an important distinction between those two terms that should cause you to raise an eyebrow as you read through some of these proposed "principles".

Single-responsibility principle: "Every class should have only one responsibility"

This principle advocates for breaking classes down into smaller pieces until each is doing just "one" thing. But there is of course nothing inherently wrong with something that does two things, particularly if those two things are always done together.

Open–closed principle: "Classes should be open for extension, but closed for modification"

This principle advocates for using polymorphism to add functionality to a pre-existing class rather than editing the class itself. There are some cases when this approach makes sense, for example at the time of writing this, I work on a team that uses the Unreal engine. We intentionally extend the Unreal-defined classes when we want functionality to change, and we try to limit engine changes. These practices are so that when we do engine upgrades, we don't have to reconcile a ton of our changes to their code with their updates to their code. But applied generally, this principle is rarely ideal. If you own the code, then just change the code. When changing code, most studios encourage engineers to make the "minimum change necessary", which is probably a better principle to follow.

Liskov substitution principle: "Base class instances should be replaceable with derived classes"

This principle focuses on maintaining intuitive code which is always a good idea.

Interface segregation principle: "Too many interfaces is better than too few"

This principle is generally understood to not be well labeled as a principle. In many cases it makes sense to break apart your interfaces, and in other cases it does not.

Dependency inversion principle: "Depend upon abstractions, not concrete instances."

The main pitfall of this pattern is premature abstraction. We can almost always make code more extensible. But generally, if we can simplify the architecture, we can achieve a more performant and data-focused approach.

GAME PROGRAMMING PATTERNS

Game Programming Patterns is a book, by Robert Nystrom, that summarizes many patterns from game development [2]. All of the chapters are available online with sample code and detailed explanations. Here I'll provide a high-level summary of each pattern he explores, but checking out his book directly is the best way to learn about these concepts if you have not heard of them before.

Double Buffer: Prevent a reader from outpacing a writer by having two buffers, "current" and "next". Perform reads on "current" and writes to "next".

Game Loop: Decouple the progression of game time from user input and processor speed.

Update Method: Simulate independent objects by telling each to process one frame at a time.

Bytecode: A series of custom byte-length instructions, used for intermediate representation in between data and code.

Subclass Sandbox: The base class defines a sandbox method, without an implementation, and helper functions, with implementations. Subclasses then need to implement the sandbox method, and they can do so using the provided helper functions.

Template Method: Base class defines a working normal template method and lets derived classes fill in specific details by overriding protected virtual methods. This is very similar to the subclass sandbox, the difference is whether the base or derived class implements the functions.

Type Object: Each instance represents a different type of object (through polymorphism) rather than using enum ranges that are fixed at runtime.

Component: Entities are simple containers of components. Each component handles one responsibility.

Event Queue: Decouple the sender from the receiver, both statically and in time.

Service Locator: A global point of access to services without coupling users to the concrete classes that implements them. This is sometimes framed as an alternative to having many globally accessible singleton services. This pattern doesn't affect the design of the service class itself, just how the service is accessed. You can send a "null" service, which doesn't do anything. And you can create decorator services that wrap default functionality to do something in addition such as a log decorator.

Data Locality: Organize data for cache coherency.

Spatial Partition: Store data structures organized by their world-space positions.

Dirty Flag: Avoid unnecessary work by deferring it until the result is needed.

Object Pool: Reuse objects from a fixed pool instead of allocating and freeing them individually.

Flyweight: Separate an object's state into two categories based on if the state will be shared between instances of the object. The shared "intrinsic" state is then stored as an immutable flyweight object that can be used by many instances.

Factory: A static method to replace an object's constructor. They can return pre-existing instances, new instances, or even subclasses. For example, let's say that Puppy is-a Dog and overrides the Adopt function. Now if we call Puppy.Adopt it will return us a Puppy, which may be new, or it could be pre-owned and just lying in an object pool. Factories only return one object, though they may support several object types. Factories can provide dynamic functions as an alternative to polymorphism, for example my class can have a function pointer Bark which can be set to point to PuppyBark or BigDogBark by the factory.

Abstract Factory: An object (not just a method) whose sole purpose is object creation, it creates a family of objects (not just one) that depend on each other. It is one higher level of abstraction than a regular factory, and often uses a regular factory. For example, we might have a DogFamilyBuilder will use factories to setup myPup = Puppy.Adopt() and myDog = Dog.Adopt(). Then it can call myPup.SetMother(myDog) to build a maternal relationship before providing both as a family.

Builder: The builder pattern is used to create complex objects with parts that must be created in the same order or using a specific algorithm. The builder has "Step1()", "Step2()", etc. methods and often an external class, the "director", which calls the builder methods.

Prototype: Instantiates a new object by copying an existing one. Can create deep or shallow copies.

Singleton: Ensures that only one object of a particular class is ever created.

Adapter: The adapter pattern is used to provide a link between two otherwise incompatible types by wrapping the "adaptee" with an adapter class that implements an interface required by the client.

Bridge: Decouples abstract elements of a class from the implementation details, providing the means to replace the implementation details without modifying the abstraction. For example, Vehicle has an Engine pointer which could point to an instance of GasPoweredEngine OR NuclearPoweredEngine.

Composite: A composite object is a class within a tree hierarchy (not strictly an is-a relationship) that has at least one child. In other words, composites are non-leaf nodes. The composite pattern says composite nodes must be accessible and utilizable in the same way as leaves. For example, we can ask a shopping bag for its "price", and it'll sum the price of all groceries it contains.

Decorator: Extend or alter the functionality of objects at run-time by wrapping them in an object of a decorator class.

Façade: The facade pattern is used to define a simplified interface to a more complex subsystem.

Proxy: Controls and manages access to the object it is protecting. For example, an internet proxy may only allow connections to certain websites on the internet.

Chain of Responsibility: Handlers can receive a request then decide if they handle it themselves or pass the request onto the next handler in a chain.

Command: A request object, with parameters, which may then be executed immediately or held for later.

Interpreter: Uses OOP to represent a language or instructions. First define a grammar, then parse input into a structure based on the grammar, and finally interpret the instruction.

Iterator: Traverses a collection of items without the need to understand its structure.

Mediator: Instead of classes communicating directly, the classes send messages via a mediator object.

Memento: Stores current state of object, so it can be restored later w/o breaking rules of encapsulation.

Observer: Observer objects subscribe to subjects to be notified if they do something.

State: Alters the behavior of an object to apparently change at run-time based on current state.

Strategy: Creates an interchangeable family of algorithms from which one is chosen at run-time.

Visitor: Separate a complex set of structured data classes from the functionality of working with their data.

GAME DESIGN

Gameplay engineers (GPEs) have one of the most versatile roles within game development. It is very common that GPEs will be asked to solve design problems. The best GPEs are often decent game designers. Here are some patterns and principles you should be familiar with for game design.

Fail Fast: Failing fast is identifying that we can spend a lot of time to build a perfect solution to the wrong problem. In game development, it's often better to find a partial solution quickly than to spend more time on the ideal solution. Students often learn about the tradeoffs between algorithms with respect to the time and space they take to run. But the largest expense developers need to be aware of is usually the time it takes to implement a particular solution. When working in a team, you should always be thinking of how quickly you can unblock your teammates so that they can make meaningful progress towards their objectives. Sometimes this approach is phrased as "minimum viable product" (MVP) where the goal is to achieve an MVP first, and then focus on polishing it once we are sure it's the right answer to our problem.

Figure 5.1 illustrates the pattern of implementing MVPs. It shows that if we want to support a player getting from place to place, building out a minimum solution, like a skateboard, can provide the player affordance quickly. This allows our teammates to progress their work as we incrementally work towards our ideal solution in incremental steps that each provide some added benefit. Of course, sometimes the MVP is the ideal solution, and we have to take the top row approach. But usually, we can deliver a working prototype along the way towards our ideal solution.

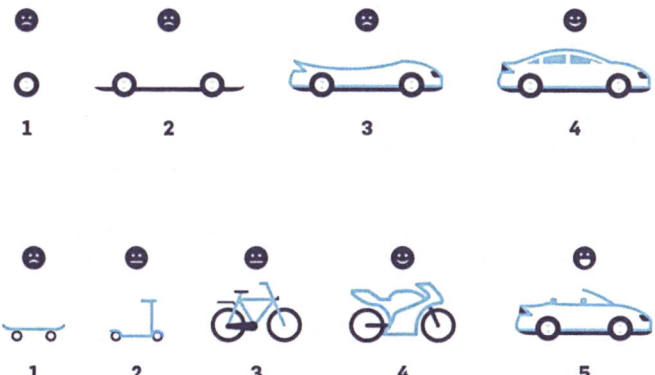

FIGURE 5.1 The top row illustrates pursuit towards an ideal solution directly. The bottom row illustrates an MVP followed by incremental changes along the way to an ideal solution [3].

Token Systems: Limit how many AI can do a thing. For example, a game may only allow three AI to attack you at once in order to prevent the player from feeling overwhelmed. This is modeled by requiring AI to acquire a token before they can attack, then having the AI release the token to allow another AI to use it for their attack.

Cooldowns: Limit how frequently an AI can do a thing. They are implemented with a simple timer that counts down a cool down period. For example, it can start when an AI attacks, then count down two seconds to prevent an AI from attacking more than once every two seconds.

Round Robin Updates: Spread updates across multiple AI to allow more AI to be supported. Imagine you have 100 AI. Instead of updating all 100 on each AI update, you can instead only update some of your AI each time. For example, on your first update, only update AI that are numbered 1-25, then 26-50, then 51-75, and finally on your fourth update you can do the last 76-100. This would increase your AI update method's speed by four times. Of course, since AI are only getting one fourth of the updates, their reaction time might slow down as well but this is a tradeoff you can negotiate.

Level of Detail (LOD): Segmented fidelity based on proximity to player. In graphics, the further something is, the less detail it needs to be rendered because a player cannot perceive all of the details from far away. The same pattern can be used for AI. If an AI is far away, hidden, or otherwise dormant, you can decrease the LOD of their thinking to make them more performant. For example, a faraway AI does not need to calculate expensive logic to test if they can melee attack you.

AI Update: When creating an AI system update, it's important to divorce it from the frame's graphical update. Typically, you will have a separate update rate for the graphics, the physics, and the AI think update. Each of these has a different required speed that can be tuned by developers.

CODE DESIGN

Within the actual implementation of code, there are many patterns you can implement to improve code's readability and extensibility. When designing software, as a general paradigm, you want to reduce state and maximize determinism. For any particular line of code, that means minimizing the ways for execution to reach that code and reducing the complexity of state present at that code. Here are a few ways this can be achieved.

Early Outs: Rather than starting an if-statement scope, return early for edge cases. Check out how the good example below reduces indentation for the damage-dealing code. This makes it easier to read, debug, and extend.

BAD EXAMPLE	GOOD EXAMPLE
| ```
void DoDamage(){
   if(PlayerIsClose())
   {
      if(WeaponIsReady())
      {
         // Deal damage
      }
   }
}
``` | ```
void DoDamage(){
   if(!PlayerIsClose())
      return;
   if(!WeaponIsReady())
      return;
   // Deal damage
}
``` |

Arbitrary Scopes: Restricts the scope of code while keeping it in-line and preventing it from being called by other methods. Generally, we don't want to early-optimize code or early-abstract code because this work introduces complexity that may not be necessary. If it's only going to be used in one place, a method like Aim() may be an early-abstraction that we can instead inline, as shown in the good example.

BAD EXAMPLE	GOOD EXAMPLE

```
void DoAttack()          void DoAttack(){
{                            // Aim
    Aim();                   {
    Fire();                      // [Aim code here]
}                            }
void Aim(){                  // Fire
    // [Aim code here]       {
}                                // [Fire code here]
void Fire(){                 }
    // [Fire code here]  }
}
```

SPECIFIC ENGINE-BASED PATTERNS

To the extent you can, try to familiarize yourself with the development environment of the team you are applying to. For example, many Unreal Engine (UE) studios require gameplay engineers to have a working knowledge of UE's Gameplay Ability System in order to be considered as candidates. The best way to prepare for the team is to see what development patterns and tools they list in the job description or to see if their team has published any GDC talks on their workflows.

WORKS CITED

[1] M. Acton, "CppCon 2014: Mike Acton 'Data-Oriented Design and C++'," *YouTube*, Sep. 22, 2014. [Online]. Available: https://www.youtube.com/watch?v=rX0ItVEVjHc&ab_channel=CppCon. [Accessed: Aug. 13, 2024].
[2] R. Nystrom, "Game Programming Patterns," *Game Programming Patterns*. [Online]. Available: https://gameprogrammingpatterns.com/. [Accessed: Aug. 13, 2024].
[3] Aldis, "Vehicle Evolution - Vehicle Evolution Diagram," *CleanPNG*. [Online]. Available: https://www.cleanpng.com/png-minimum-viable-product-agile-software-development-2073181/. [Accessed: Aug. 25, 2024].

Multithreading

6

AVOIDING SYNCHRONIZATION STRUCTURES

Similar to patterns, synchronization structures (such as locks and semaphores) are sometimes interpreted by students as the solution to multithreading problems when they are often actually the problem themselves. Your goal should be to understand these structures but then, in practice, you should strive to design code that does not need them.

> Multithreaded programming is about how to design programs to not need locks.

To emphasize how bad locks are, let's review some numbers from the previous chapters. We discussed the great penalty incurred by cache misses, which may waste 10-20 cycles on an L1 cache miss or almost ten times that amount on an L3 cache miss. And while we spent a lot of time discussing how to avoid cache misses... even if you miss every cache, and have to read from main memory itself, that cost is dwarfed by the cost of a context switch. A context switch can easily cost 10x to 100x the cost of a main memory read. We are talking about thousands of cycles lost. Context switches are terrible. And we risk a context switch every time we wait on a lock. We really need to avoid that from happening. In this chapter we will explore a number of synchronization structures, review how to use them, and then discuss strategies on how to avoid them.

SYNCHRONIZATION MECHANISMS

Mutex

A "mutual exclusion locking mechanism" (it waits if access not granted).

- It may be helpful to think of a mutex as not a 'lock' itself, but instead as an object (like a door) that you lock and unlock. You can actually have multiple

 DOI: 10.1201/9781003565550-7

locks objects operate on one mutex such as when you want a read-lock and write-lock, this is called a shared mutex.

- Recursive mutexes can be locked multiple times and must be unlocked the same number of times.
- If a thread tries to lock a non-recursive mutex that it already owns, it will deadlock.
- The call to lock() is blocking, meaning that it only returns when it has the lock. An alternative method, trylock(), returns immediately whether or not it succeeds in obtaining the lock.
- When a thread is blocked by a lock, it cannot do anything. It's at this point that the processor may do an expensive "context switch" which includes recording and replacing register values and then allowing a different thread to start work on the processor.

Lock Types

Here are some common lock types you should familiarize yourself with for the interview. Many of them have multiple names.

- **Write / "unique" / "exclusive" lock**: gives a process exclusive access for writing. While a write lock is in place, no other process can lock that part of the file for reading or writing.
- **Read / "shared" / lock**: prohibits another process from requesting a write lock on the shared resource while allowing other processes to request additional read locks.
- **Lock-guard / scoped lock**: an exclusive lock that locks on construction, then unlocks in upon destruction (an RAII approach to locks).
  - Unlike shared/exclusive locks, it <u>only</u> locks/unlocks during its construction/destruction.
- **Fine-grain locks / "striped locks"**: divide the control of an object across subdomains, each with their own lock, such as subarrays of an array.
- **Spinlock**: a design alternative to a "true" locking mechanism. Rather than wait like a normal lock, a spinlock will just continually repeat a locking request.
- They are very expensive at a high level, but a mutex is actually often implemented using a spinlock to manage its internals.
- Sometimes it's better to use a spinlock than a mutex, for example if multiple threads are making quick edits to the same object, it would be better to have one trying again and again for access to the array (which should be granted soon) than to have a thread wait and risk a context switch. Don't use a spinlock if the wait is long.

Practice Problem 6.1: How would you optimize a structure for many readers but few writers?

Solution: You can use a shared mutex that differentiates assessor privileges. Basically, this would mean two locks, one for reading and one for writing.

Condition Variable

The condition variable (CV) can be thought of as a "mutex with notifications". It's a synchronization primitive that can be used to block a thread, or multiple threads at the same time. It remains waiting until another thread modifies a shared variable, called the "condition", and then notifies the condition variable.

Example usage: Imagine you have a producer-consumer problem, where one or more threads (producers) are adding data to a shared queue, and other threads (consumers) are processing (or removing) this data. A consumer thread would call Wait() on a condition variable when the queue is empty, waiting for the producer to add more data. Once the producer adds data, it signals the condition variable, and the waiting consumer thread can wake up and start processing the data.

When a CV performs its Wait() function, it first unlocks the mutex, this allows the data it is waiting for to be changed. In terms of the earlier example, the consumer would unlock the shared queue so that producers can add to it.

Here's what the example consumer implementation might look like.

```
std::unique_lock<std::mutex> lock(mtx);
while (queue.empty()) {
    cv.wait(lock);
}
// Handle new queue entries
```

Notice that the wait function is contained within a while loop. That means it will only enter into that loop if there is nothing in the queue to consume, but it importantly also means that it will not leave the while loop while that true. This is important because CVs experience "spurious wakeup" which means that they sometimes finish their waiting before they are actually notified. Spurious wakeups are a concession of optimization, as they reduce the amount of information a kernel needs to keep track of for each thread. But we need to be aware of this behavior and always design CVs like in the code above so that they will go back to waiting if they wake up early.

Some additional notes on CVs:

- CVs allow you to wait until an arbitrary condition has changed, whereas a mutex only allows you to wait until the lock is available.
- CVs can be notified via notify_one or notify_all which determines if one or all waiting threads are awoken. Notify_one is useful if you know that only one waiting thread is going to make progress and therefore you don't need to wake all of them.
- The default CV requires a unique_lock, however CV_any can be used with any lockable object.

Semaphore

A semaphore is like a CV that uses an atomic count instead of a mutex. Like a CV, a semaphore relies on signaling mechanisms to manage access to resources and control concurrency. It uses an atomic count to govern how many threads can access a resource simultaneously.

The main types of semaphore to be aware of are "counting" and "binary" semaphores. Basically, they work the same, the only difference is that binary semaphores only count up to one.

- "Counting" Semaphores
  - They have node value restriction, besides maybe int_max.
- "Binary" Semaphores
  - Their value is restricted to zero and one.
  - The wait operation only works when the semaphore is one.
  - The signal operation succeeds when semaphore is zero.

A binary semaphore can be used just like a mutex, however a mutex cannot be used as a semaphore. That's because the mutex can only be signaled by the thread that called the wait function, whereas a semaphore can be signaled from other threads.

A semaphore API has two main functions, V and P. These were named in Dutch (by Dijkstra himself!), so the translations, Verhogen and Proberen are not very helpful for English readers. So, we usually refer to them as Up and Down, respectively.

Up / V / "Signal": Increments the semaphore's count. This action can release a blocked thread if the semaphore's count was zero before the increment.

Down / P / "Wait": If the semaphore's count is above zero, it decrements the count and moves on. Otherwise, it waits until the count is above zero.

Practice Problem 6.2: Imagine a snack machine where the vendor can add snacks when it's running low. Buyers try to take snacks from the machine and will wait if the machine is empty. How can you implement this relationship if the vendor and buyers represent multiple threads?

Solution: Use a semaphore to track the number of available snacks. Initialize the semaphore with the number of snacks available. When the vendor adds a snack, increment the semaphore count. The vendor can wait and be notified when no more snacks are available. When a buyer takes a snack, they will decrement the count. If there are no snacks to buy, the buyer will wait and can be notified when snacks are available.

ATOMICITY

Atomic operations are completed before another process can interrupt. They can provide us with an alternative to some of the synchronization structures (like mutexes) that we previously discussed.

For example, if we have two threads that are incrementing a value, then we need to have a lock so that the threads can take turns obtaining ownership of the shared data before editing it. However, if we can ensure that our increment is atomic then we don't need the lock because by the time the second thread wants to edit the data, the first thread will be done editing it.

An operation is atomic if the data is aligned, and the bus is at least as wide as the type being written or read. This means that on x64, there is no guarantee that reads and writes larger than eight bytes will be atomic.

Practice Problem 6.3: Read through each of example operations [1]. Which ones are atomic?

```
== A ==
DWORD* pData = (DWORD*)(pChar + 1);
*pData = 0;

== B ==
int g_globalCounter;
++g_globalCounter;

== C ==
std::atomic<int> x;
x++;

== D ==
int g_alignedGlobal;
g_alignedGlobal = 0;
```

Solution: A is not an atomic write because the pData variable is not natively aligned.

B is not atomic because the ++ operation is several operations in a single statement. Neither x++ nor ++x will be atomic by default. If we wanted this increment to be atomic, we would need to do one of a few concurrency strategies. Such as replacing the int type with std::atomic, or we could add a lock around the increment, or we could use an intrinsic function that guarantees atomicity such as InterlockedExchangeAdd. The note that this variable is a global variable is irrelevant to its atomicity.

C is atomic. The variable is an atomic type so the ++ operator is implemented to guarantee atomicity.

D is probably atomic. The variable is aligned, and it is an int which should be less than the processor's native word size. These are both requirements for atomicity. However, they do not guarantee atomicity. It is ultimately architecture-dependent, but if we assume that we are talking about an x86 or x64 architecture, then this is atomic.

ATOMIC OPERATIONS

Which operations are atomic will vary based on the hardware you are running on. But typically, you will have access to a few atomic APIs such as the ones below. These are implemented thanks to hardware support. All of these functions will be sequentially consistent (a topic we will discuss in a future section on fences).

- **Fetch-and-Add (FAA)**: Increments the value at address and returns the original value that was stored at the address.
- **Fetch-and-Subtract (FAS)**: Same as FAA, but with subtraction.
- **Atomic-Exchange**: An atomic setter. You "exchange" a variable's current value for a new one which sets the variable to the new value and returns the old value.
- **Interlocked-Increment**: Basically just an atomic version of i++.
- **Compare-and-Swap (CAS)**:
  - Takes in an address, a new value, and a comparison value.
  - If the value found at the address equals the comparison value, CAS will set it to the new value.
  - Returns true if the swap was successful.

Practice Problem 6.4: Imagine a linked-list that uses a lockless CAS approach to update nodes. Several threads, T1, T2, and T3, will try to update the list.

- T1 will try to remove node A, which holds a value of 77.
- T2 will try to add node B, with a value of 77.
- T3 will try to replace node A from its current value to 88.

Here's how the scenario plays out. First, T3 navigates to node A and detects that node A has a current value of 77. It prepares a CAS operation with the following parameters:

- Address to change: node A
- Comparison value (value to replace): 77
- New value: 88

Before T3 invokes the CAS, T1, then T2, successfully complete their objectives. In this scenario, nodes are pooled. So, when T1 removes node A, it returns to a pool and is reused as node B for T2.

Now that T1 and T2 are finished, T3 is able to execute its CAS which is performed successfully… or is it? Actually, T3 has mistakenly replaced node B's value! Since that memory is now used for node B and just happened to have the matching comparison value of 77. How could we have avoided this?

> **Solution**: This question highlights a major weakness of CAS and is famous enough to have its own name, the ABA problem. Ultimately, CAS alone cannot safely solve this problem. There's always a chance that the comparison value might be coincidentally set at the targeted address despite some in-between work, called "B", having invalidated the CAS objective. The most common fix is to add more data so that we can detect if node A is still the node A we wanted to replace. For example, one approach here might be to give nodes a version number in addition to their value. So that when T3 first sees A, A will have value "77, v1". Then when T3 checks B's value during the CAS, T3 will see the value "77, v2". The difference in versions will allow T3's CAS to detect that the value has changed at this memory address and therefore that it should abort.

ATOMIC TYPES

Atomic types are types that promise to perform all or a subset of their operations as atomic operations. Atomic types are not necessarily lock free. Atomic types contain "critical sections" that must be performed atomically and that usually block system interrupts as well.

- std::atomic<T>::is_lock_free() can be used to check if an atomic has a lock-free implementation

- Beware that many atomics are locking. This means that objects which use them (such as Shared_Pointers) may lock.

Atomics vs Locks

In this section we'll compare these two types of objects. For this discussion, we assume that "atomics" refers to non-locking atomics.

- Locks are easier to think about; and we can context switch to be useful while waiting, whereas non-locking atomics just busy wait aka "spin-lock".
- To make lock-free data structures, first remove locks, and then re-design the data flow to use non-locking, atomics.

Atomics:
- They leverage processor operations, like CAS, to not use locks at all.
- They don't wait, instead they keep trying until success, sometimes called busy waiting.
- They don't incur context-switching overhead, but they also don't free up CPU resources.
- They're faster if contention between threads is sufficiently low.
- They don't cause a thread to stop others from making progress.

Locks:
- They're more OS-dependent and perform differently on, for example, Win and Linux.
- They may prevent other threads from making progress.
- They suspend thread execution.
- They free up CPU resources for other tasks.
- They incur context-switching overhead when stopping/restarting the thread.
- With locks, we can make a tiered system of readers and writers.
- They're usually much simpler to implement and think about.

Neither approach is clearly better, context matters. I recommend Travis Down's blog post titled "A Concurrency Cost Hierarchy" for a breakdown, with performance metrics, on several approaches [2].

MULTITHREADING DATA STRUCTURES

In a previous chapter we explored common data structures and their tradeoffs. These structures can be converted into multithread compliance by simply "putting a lock on it", or in other words requiring a thread to acquire a lock before accessing the data.

But many of these structures can also be converted into lockless multithreaded structures, also known as non-blocking structures. In this section we will review the high-level steps of a lockless list's Insert and Delete.

Lockless List Insertion

Consider a request to add newElem right after curElem. First, we walk up the list until we find curElem and its next element "curElem.next". Then we set up our new element to point to that "next" value. That's because once we put newElem in front of curElem, the element that was previously after curElem will now come after newElem. We need to set this before the CAS because once we need to make sure we leave the list in a working state after we exit our critical section of changes. Here are the steps in order:

1. next ← curElem.next.
2. newElem.next ← next.
3. CAS(&curElem.next, next, newElem).
4. If the CAS was not successful, go back to Step 1.

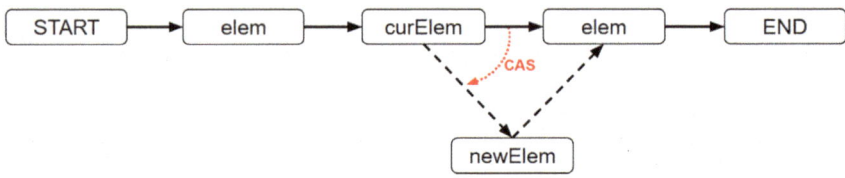

FIGURE 6.1   Diagram of lockless insertion.

Lockless List Deletion

Consider a request to remove staleElem from a list. For deletion, a single CAS operation is not sufficient [3]. That's because CAS can only swap one value at a time, so if you change the element before staleElem to point to the node after staleElem, there's a possibility that during that change, another element may have been added after staleElem. This example is illustrated in Figure 6.2, where START previously pointed at staleElem, and START was CAS'd to point at elem instead, but while that was happening newElem was added after staleElem. The result is that when we delete staleElem we will leave newElem hanging.

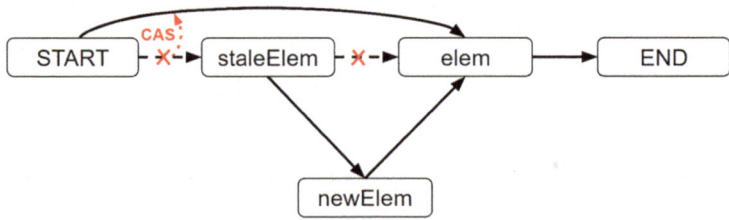

FIGURE 6.2   Diagram of what can happen if only one CAS is used in a lockless list deletion.

One correct approach is to perform the deletion in two passes. In the first pass, we mark a node as "logically" deleted. This means we mark it so that new nodes will not be placed after it. Then we can safely return to the lost and perform the actual deletion.

1. Find the node to delete.
2. Mark it as "logically" deleted (usually by using CAS to set some marker on the node), so that future insertions will not insert right after it.
3. Actually delete the node using a second CAS.

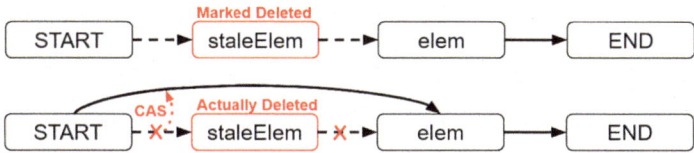

FIGURE 6.3 Diagram of lockless deletion.

MISCELLANEOUS MULTITHREADING

False Sharing: A performance issue in multi-threaded programs where threads unintentionally access different variables that share the same cache line, leading to unnecessary cache invalidations. This causes the CPU to waste time updating the cache, resulting in slower performance. The problem can be avoided by adding padding or aligning data so that each thread's variables are stored in separate cache lines.

WORKS CITED

[1] Ali_Bahoo, "Can atomic operations prevent context switches?," *Stack Overflow*, Dec. 27, 2013. [Online]. Available: https://stackoverflow.com/questions/20858395/can-atomic-operations-prevent-context-switches. [Accessed: Aug. 13, 2024].

[2] T. Downs, "A Concurrency Cost Hierarchy," *Performance Matters*, July 6, 2020. [Online]. Available: https://travisdowns.github.io/blog/2020/07/06/concurrency-costs.html. [Accessed: Aug. 13, 2024].

[3] T. L. Harris, "A Pragmatic Implementation of Non-Blocking Linked-Lists," in *Proceedings of the 15th International Symposium on Distributed Computing*, 2001. [Online]. Available: https://www.cl.cam.ac.uk/research/srg/netos/papers/2001-caslists.pdf

PART II

Maths

Maths
Vectors

7

Maths is certainly the topic that typical software engineers have the most trouble with when applying to game development engineering jobs. Game devs, particularly gameplay programmers, rely on 3D maths to solve 3D problems. In this chapter we will cover vectors, the foundation structure for most of the maths we will be doing. And we'll cover the most common topic asked on the gameplay programming interview: the dot product.

INTRODUCING VECTORS

Quick Reference for Notation

Vectors are typed in bold lowercase, for example: $\mathbf{v}$.

Unit vectors are typed in bold lowercase italics with a hat, for example: $\hat{\boldsymbol{v}}$.

Matrices are typed in bold uppercase, for example: $\mathbf{M}$.

Points are typed in uppercase but not bolded, for example: P.

Lines and line segments are typed in terms of the points they contain, for example: P_1P_2.

Scalar variables are typed in lowercase but not bolded, for example: x.

Structure like planes, as a single variable, are typed in bold uppercase italics, for example: $\boldsymbol{P}$.

Some Notes on Notation

A vector is a data structure that holds numbers. A vector is like a matrix, but a vector is always only one column or row. In that sense, a vector is like a very tall or very wide matrix. Matrices are described as m×n, pronounced "m by n", where m is the row count and n is the column count. When typed, a vector (like $\mathbf{v}$) will always refer to a tall, vertical, one-column vector. In terms of m×n this means n = 1. If we want to refer to a wide

DOI: 10.1201/9781003565550-9

vector, where m = 1, then we write $\mathbf{v}^T$, this looks like "$\mathbf{v}$ to the power of T" but is simply pronounced "$\mathbf{v}$ transpose".

$$\text{If } \mathbf{v} = \begin{bmatrix} 1 \\ 2 \\ 3 \end{bmatrix}, \text{ then } \mathbf{v}^T = \begin{bmatrix} 1, 2, 3 \end{bmatrix}.$$

If I write $\mathbf{v} = [1, 2, 3]$, then I am always describing a vertical vector that looks exactly like the $\mathbf{v}$ shown on the line above but I am just writing it horizontally for typing practicality. If a vector is actually horizontal (wide) then I will refer to it using the transpose notation.

The zero vector, written as $\mathbf{0}$, is a vector of all zeroes. The number of entries in the zero vector is implied based on context.

Rays, Lines, Line Segment, and Vector Formulas

A line is a 2D structure that has zero width, zero height, and infinite length. In school you may have learned the line equation y = mx + b. We will not use this equation; it is a very constrained 2D representation of a line that will not be useful for game dev where most of our applications are in 3D.

Instead, I want you to think of a line as a container in 3D space. Our line formula is an equation that defines which points are on our line. In that regard, a line equation can be thought of as a test to see if any given point is on our line.

A line contains an infinite number of points, but two points are sufficient to define the line. If we have two points, P_1 and P_2, we can subtract them to get the direction between them, $\mathbf{v}$. The line containing P_1 and P_2 can be defined using this $\mathbf{v}$ in addition to one of the points on the line. The point on the line we choose is represented as P_n (it can be P_1, P_2, or any other point on the line).

$$P_1P_2 = P_n + (t * \mathbf{v})$$

The formula above defines line P_1P_2, starting from point P_n, into the direction $\mathbf{v}$ which is scaled by scalar t. Because a line extends in two directions, t can be negative. Because a line is infinite, t can be any value from -infinity to +infinity. Now here is the cool part... this is also the equation for a direction vector and a line segment. The main difference between a vector, line, and line segment is the range of our scalar, t. Below are the parametric equations for each of these structures, so-called "parametric" because they all rely on that input parameter t.

$$P_1P_2 = P_n + (t * \mathbf{v}) \text{ is a ray if } t = [0, +\text{infinity})$$
$$P_1P_2 = P_n + (t * \mathbf{v}) \text{ is a line segment if } t = [0, n]$$
$$P_1P_2 = P_n + (t * \mathbf{v}) \text{ is a line if } t = (-\text{infinity}, +\text{infinity})$$

The realization that we can sometimes think of rays and line segments as "lines with bounds" is very helpful. Most of the problems that we will face with these types are much easier if we instead solve the problem with lines. For example, we will later explore the problem: "does line segment P_1P_2 intersect plane P?", the solution is basically to first solve "does <u>line</u> P_1P_2 intersect plane P?" and if it does, then we just need to check if the intersection fits within the bounds of line segment P_1P_2. I call the two steps of this method solving the "ideal" and "real" problems.

Below is the formula for a vector. You may have noticed that the vector has no P_n component in its definition. That's because vectors have a magnitude and a direction but no position. You may freely move a vector, and it will remain the same vector.

$$P_1P_2 = t * v \text{ is a vector if } t = \left[-\text{infinity}, +\text{infinity} \right]$$

Practice Problem 7.1: What is the equation of a line that contains these two points: (11, 13, 17) and (10, 11, 14)?

Solution: First, define the direction vector of the line which is the difference of two points within the line.

$$v = \left(11, 13, 17\right) - \left(10, 11, 14\right) = \left[1, 2, 3\right]$$

Then define the line as a starting point plus the direction vector scaled by t (a variable scalar).

$$P_1P_2 = P_n + \left(t\, v\right) \rightarrow P_1P_2 = \left(11, 13, 17\right) + t * \left[1, 2, 3\right].$$

3D Shapes from 2D Vectors

We discussed how a line's equation is like a formula to see if a point is on a line. Abstracting that idea a bit... a group of vectors, when considered together, can be thought of as a formula as well. For example, a plane is a 2D structure that can be defined by two vectors. Any two vectors will work as long as they are not coincident aka collinear aka in the same direction. We'll explore planes in far greater depth later, but the key insight here is that a line or vector can be thought of as formula of points and so too can a plane be thought of as an equation of vectors. Now let's take that one step further into 3D...

Basis Vectors

A basis is a collection of vectors that can be combined in order to reach any point within a space. Here's an example:

$$a\mathbf{v} + b\mathbf{u} = P$$

If 2D vectors $\mathbf{v}$ and $\mathbf{u}$ can be combined and scaled to express any 2D point P, then $\mathbf{v}$ and $\mathbf{u}$ form a basis for R^2 (which means 2D space).

- For example, let's say P = (1, 10), $\mathbf{v}$ = [1, 0], and $\mathbf{u}$ = [0, 1]. Then the earlier equation can be completed with a = 1 and b = 10.
- But if P = (1, 10), $\mathbf{v}$ = [1, 0], and $\mathbf{u}$ = [2, 0]. Then the equation can never be completed because the second term of P, 10, can never satisfy the equation a*0 + b*0 = 10.

To form a basis for a vector space, you need the same number of vectors as the dimension of the space, and you need all of them to be disjoint aka non-colinear aka not in the same direction. I like to think of basis vectors as rigid segments of a robotic arm. It doesn't help if two segments go the same direction, in order for the robot to be able to reach any point, the segments need to extend in different directions.

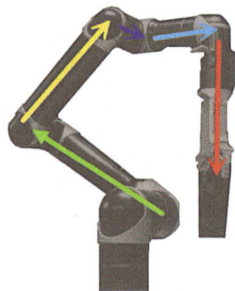

FIGURE 7.1   A robotic arm [1].

Practice Problem 7.2: Which of the following are a basis in their respective dimensions?

- Three disjoint vectors in R^4.
- Two collinear vectors in R^3.
- Vectors [0, 1], [0, 2], and [5, 1].

Solution: Three disjoint vectors in R^4: These vectors are disjoint, that's important! But we need four to form a basis in R^4.

Two collinear vectors in R^3: These vectors are on the same line; therefore, they alone cannot form a basis. Additionally, we would need at least three vectors to form a basis in R^3.

Vectors [0, 1], [0, 2], and [5, 1]: Here we have three vectors in R^2. The first two are colinear but the third is not. So, if we take the third and either of the first two, we have two disjoint vectors in R^2 which is sufficient to form a basis.

Bases are really important in game development. An object's "transform" is a basis that defines its orientation. Have you ever seen an image like Figure 7.2? This is commonly how transforms are visualized. As you can see, there are three disjoint vectors shown. All of a transform's basis vectors are perpendicular, aligning with the word's x, y, and z axes. Transforms are always drawn with the x, y, and z (XYZ) axes colored red, green, and blue (RGB) respectively.

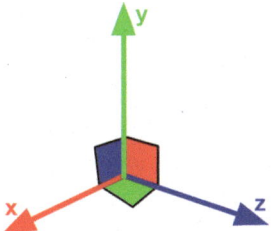

FIGURE 7.2   A transform.

Vectors for Distance

Vectors are commonly used to define the space between objects. In an earlier example we discussed you can take two points on a line, P_1 and P_2, and subtract them to derive the direction of the line. The order of the points in the subtraction (P_1-P_2 vs P_2-P_1) doesn't matter for a line because it is infinite in both directions, but it matters a lot when directionality is relevant. For example, if S is a race's start position and E is the race's end position, how do you derive the direction from the start to the end of the race? It's E–S. You always put the end point as the first argument. A mnemonic for this is "think of the destination first." If you ever forget, it's easy to derive the correct order by using a 1D example. Just consider if you want to get from 0 to 10, do you need "10–0 = +10" or "0–10 = -10"? Of course we need the destination of 10 to come first, otherwise the result is negative.

Vector Norms

The magnitude of a vector is the total distance it represents. The vector magnitude is also known as the "L2-norm", "Euclidean norm", or simply "the norm". A norm is usually expressed as one or two lines on each side of a vector, "‖v‖". A vector norm is like an absolute value formula for vectors. In fact, for a 1D, meaning single-element vector, the norm is the absolute value of its one value. But for higher dimensional vectors, we have a formula to help us out:

$$\|v\| = \text{sqrt}\big((\, v.x * \; v.x \,) + (\, v.y * v.y \,) + \ldots\big)$$

This norm (the L-2 norm) is the most common kind of norm which is why people sometimes just call it "the norm". But there are actually many kinds of norms. The L-2 norm belongs to a family of norms called p-norms which vary based on the value of a variable "p". The L2-norm is simply the p-norm when $p = 2$.

Practice Problem 7.3: What is the magnitude of vector **v**, where **v** = [2, 3, 4]?

Solution:

$$\|v\| = \text{sqrt}\big((\; v.x * v.x \,) + (\; v.y * v.y \,) + (\; v.z * v.z \,)\big)$$
$$\|v\| = \text{sqrt}\big((\, 2 * 2 \,) + (\, 3 * 3 \,) + (\, 4 * 4 \,)\big)$$
$$\|v\| = \text{sqrt}\big((\, 4 \,) + (\, 9 \,) + (\, 16 \,)\big)$$
$$\|v\| = \text{sqrt}(\; 29 \;)$$

Normalizing Vectors

Often, it's helpful to "unitize" a vector, which means scale it to be of length 1. We do this sometimes when we want a vector only for its direction. To normalize a vector, divide each term by the vector's magnitude. If a vector is normalized, then it will be italicized.

$$\hat{u} = v \, / \, \|v\|$$

Practice Problem 7.4: Unitize vector **v**, where **v** = [2, 3, 4], into a unit vector $\hat{u}$.

Solution:

$\|\mathbf{v}\| = \text{sqrt}(29) \leftarrow$ From Practice Problem 7.3

$\hat{u} = \mathbf{v} / |\mathbf{v}|$

$\hat{u} = [2, 3, 4] / \text{sqrt}(29)$

$\hat{u} = [2 / \text{sqrt}(29), 3 / \text{sqrt}(29), 4 / \text{sqrt}(29)]$

DOT PRODUCT

Congratulations, you have made it to the most important topic of this entire book. The dot product aka "the inner product" is absolutely the most common interview topic out of all possible game programming questions. I have never experienced a gameplay engineering interview that has not touched upon the dot product. It is imperative that you learn this topic inside-and-out, upside-down, and sideways.

The dot product is the product of two vector magnitudes multiplied by the cosine of the angle between the vectors, shown as the left-hand side expression of the following equation:

$$\mathbf{a} \cdot \mathbf{b} = \|\mathbf{a}\| \, \|\mathbf{b}\| \, \cos(\theta)$$

Its primary uses are to find the angle between two vectors and to perform vector projections. To remember that cosine needs to be used for the dot product, I suggest the mnemonic of observing that either the 'o' or 'c' in cosine look like a dot.

The dot product can be expressed in many ways, each formula has its own uses. Here are some common expressions followed by explanations and greater context on their significance.

Standard Form

$$\mathbf{a} \cdot \mathbf{b} = \|\mathbf{a}\| \, \|\mathbf{b}\| \, \cos(\theta)$$

The most common dot product expression. You must have this memorized. Theta ("θ") represents the angle between vectors **a** and **b**. To solve for theta, we can divide both sides by ‖**a**‖ * ‖**b**‖ and then use the arccos() function on both sides to get theta alone on the right hand side.

If **a** and **b** are of unit length, aka "unit vectors", then their magnitudes are 1 and their magnitude factors (‖**a**‖ and ‖**b**‖) can be removed to simplify the formula to $\text{acos}(\mathbf{a} \cdot \mathbf{b}) = \theta$.

The theta we receive from a dot product formula is unsigned meaning it does not have any directionality. So, if you have two vectors, the dot product can tell you how much to rotate the first for it to align with the second, but it can't tell you which direction to turn. We will later explore the cross product which does provide signed directionality.

Unitized Form

$$\hat{\boldsymbol{u}} \cdot \hat{\boldsymbol{v}} = \cos(\theta)$$

If the two input vectors are unitized, meaning that they are both of unit length, then the standard form can be simplified to the unitized form above.

I have a riddle for you:

Practice Problem 7.5: If the unitized form's left-hand expression, $\hat{\boldsymbol{u}} \cdot \hat{\boldsymbol{v}}$, results in a negative number, then what does this tell us about the angle theta?

Solution: Think about the unit circle which I have included as Figure 7.3. Notice that cos(x) is only negative if the angle is in the top left or bottom left quadrant of the unit circle. Therefore, if $\hat{\boldsymbol{u}} \cdot \hat{\boldsymbol{v}}$ results in a negative number, then θ must be obtuse.

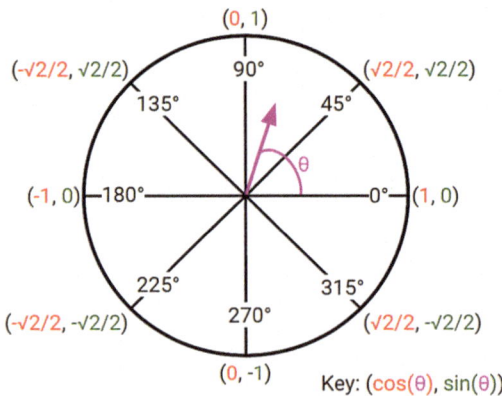

FIGURE 7.3 The unit circle. Inside, angles are represented by degrees. Outside, coordinates are represented in (x, y) format. The pink angle θ appears to be between 45° and 90°, so its x value will be between the x value of 45° (√2/2) and the x value of 90° (0). The x values represent cos(θ), and the y values represent sin(θ). For example, cos(90°) = 0. We will discuss the unit circle more in the rotations chapter of this book.

Expanded Form

$$\mathbf{a} \cdot \mathbf{b} = \mathbf{a}\ \mathbf{b}^{\mathrm{T}} = (\mathbf{a}.x * \mathbf{b}.x) + (\mathbf{a}.y * \mathbf{b}.y)$$

This formula shows that to actually compute the dot product you sum several products where each product is a pair of each vectors' components. Vectors are m×n matrices, where n = 1. Two matrices cannot be multiplied unless the n of the first factor equals the m of the second factor. So you cannot multiply two vectors because you can't multiply an m×1 by m×1, when m != 1. However, if we transpose the second vector, it becomes "1×m" and now we can multiply an m×1 x "1×m" because the 1's in the middle are equal. This is effectively what the dot product does: it transposes the second vector and then performs basic matrix multiplication. You may notice that the right-hand side of this formula looks very similar to the vector norm formula, except there is no square root and we have two distinct vectors. We will explore that similarity in the next formula.

Squared Norm Form

$$\mathbf{a} \cdot \mathbf{a} = \|\mathbf{a}\|\ \|\mathbf{a}\| = \|\mathbf{a}\|^2 = \mathrm{sqrt}\big((\mathbf{a}.x * \mathbf{a}.x) + (\mathbf{a}.y * \mathbf{a}.y)\big)2 = (\mathbf{a}.x * \mathbf{a}.x) + (\mathbf{a}.y * \mathbf{a}.y)$$

This formula shows that a vector dotted with itself is equal to its squared magnitude aka its squared norm. This basically amounts to the vector norm formula but the square cancels out the square root.

Cauchy-Schwarz Form

$$\left|\mathbf{x} \cdot \mathbf{y}\right| \le \|\mathbf{x}\|\ \|\mathbf{y}\|$$

Check out that cool triangle in Figure 7.4. It's formed by vectors $\mathbf{x}$, $\mathbf{y}$, and $\mathbf{z}$ with lengths of x, y, and z respectively. Triangles have the interesting property that |x + y| ≤ |x| + |y|. This is called the Triangle Inequality, and it means that any one side length of the triangle cannot be greater than the sum of the other two side lengths. Think about why this is true, if $\mathbf{z}$ had a magnitude greater magnitude than the sum of $\mathbf{x}$'s and $\mathbf{y}$'s magnitudes, then $\mathbf{x}$ and $\mathbf{y}$ could not be combined to go from the start of $\mathbf{z}$ to the end of $\mathbf{z}$. Similar to the Triangle Inequality of the vector magnitudes, we can express an inequality using the dot product: |x · y| ≤ ||x|| ||y||. This is known as the "Cauchy-Schwarz" inequality.

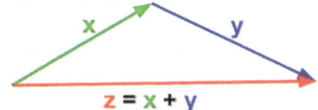

FIGURE 7.4   A triangle formed by three vectors.

Interpreting the Standard Form

$$\mathbf{x} \cdot \mathbf{y} = 0 \text{ iff } \mathbf{x} \perp \mathbf{y}$$

When you complete a dot product, if (and only if) the result is zero then that means the vectors are perpendicular aka orthogonal (regardless of if the vectors are normalized). Testing if the dot product of two vectors is zero is actually the easiest and most common way to check if vectors are perpendicular. I'd like to combine this insight with the previous equations' insights to present you with the following statements which we can use to interpret the result of any dot product.

$\mathbf{a} \cdot \mathbf{b} = \|\mathbf{a}\| \|\mathbf{b}\|$ if they're exactly same direction

$\mathbf{a} \cdot \mathbf{b} > 0$ separated by an acute angle

$\mathbf{a} \cdot \mathbf{b} = 0$ if perpendicular

$\mathbf{a} \cdot \mathbf{b} < 0$ separated by an obtuse angle

$\mathbf{a} \cdot \mathbf{b} = -1 \|\mathbf{a}\| \|\mathbf{b}\|$ if they're exactly opposite direction

Practice Problem 7.6: Imagine you have the positions of a player and an enemy, P and E. You also have their forward vectors, **p** and **e**. Determine if the enemy is facing the player.

Solution: A lot of interview questions are designed to be rather ambiguous in order to test a candidate's ability to take a problem from abstract design ideas into concrete engineering specifications. This is an example of that kind of question, there is a lot of ambiguity here. When you get a question like this, you should identify ambiguities, propose a direction based on your prior experience, and pursue a concrete solution.

In this case, you might say "I'd first work to define what 'facing' means. If I can't get a clear direction from the team, I will assume that it means the player is no more than 45 lateral degrees from the enemy's forward vector". This basically means that enemies have a 90-degree vision cone or field of view (FOV).

Now we can start working toward a solution. We can calculate the angle between **p** and **e** using the dot product and then determine if the angle is within the acceptable threshold.

You should also list edge cases for this problem and how you would approach them. For example:

- What if there is a wall between the player and enemy?
- What if the player is very far away from the enemy?
- What if the player is above the enemy?

VECTOR PROJECTIONS

Vector projection is the process of isolating a portion of a vector. Take a look at Figure 7.5 which shows four vectors: **a**, **b**, **x**, and **y**. Vector **a** can be broken into two different vectors: **x** and **y**. For example, **x** and **y** might represent the portions of **a** that go in the directions of the x-axis and y-axis respectively. In other words, **x** and **y** are component vectors of **a**. If I asked for the "projection of **a** onto **b**" then that would be the same as asking for "the component of **a** that goes in the direction of **b**". In this case that would simply be **x**, because **x** and **b** are coincident. Written in maths notation, this relationship may be expressed as $\text{proj}_b\,\mathbf{a} = \mathbf{x}$.

It may help to think of a sun projecting the shadow of **a** onto **b**. Can you imagine how **x** in Figure 7.5 is like a shadow of **a**? Some students love this analogy, but I think it has limited usefulness since projections in general can be upside down and usually rather unlike shadows. The main insight is to realize that **x** is the part of **a** that goes in the direction of **b**.

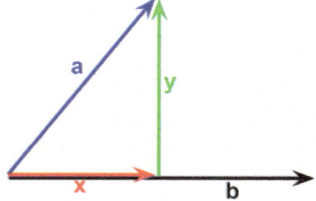

FIGURE 7.5 Four vectors, where **a** = **x** + **y**, and **x** is the projection of **a** onto **b**.

In Figure 7.6, I have moved **b**. Now the projection of **a** onto **b** is **z**. This is a really important example because it shows that the components of a vector do not need to align with the x-axis or y-axis. You can derive a vector's components in any direction.

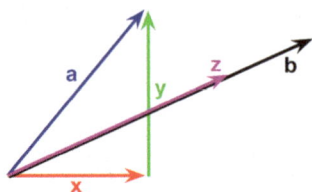

FIGURE 7.6   Four vectors, where **a** = **x** + **y**, and **z** is the projection of **a** onto **b**.

Practice Problem 7.7: In Figure 7.6, **z** is a component vector of **a**. Derive **w**, where **w** is the remaining component vector of **a**. Assume we are in a 2D context.

Hint: In a 2D context, two non-colinear components together will define a vector. That's because two non-colinear vectors form a basis in R^2 which can be used to represent any point in that space. Using this insight, we can form the following equation: **z** + **w** = **a**. Just like how **x** + **y** = **a**.

Solution: Since **w** is the remaining part of **a**, that must mean that **a** = **z** + **w**. Therefore **w** = **a** - **z**. But here's something really cool... in this 2D example, if **z** is **a**'s component in **b**'s direction, then that means **w** is **a**'s component in the direction perpendicular to **b**. So just like we were able to project **a** onto **b** to get **z**. If we found some vector perpendicular to **b**, say **c**, then the projection of **a** onto **c** would be **w**. This is a 2D case, so **a** only has two components. But if we were in 3D then we could extrapolate this method further to find a final vector, **d**, that would be perpendicular to both **b** and **c**. Projecting **a** onto **d** would give us our third component. Projections are like chopping up our vector into its composite parts.

Projection Formula

The vector projection of **a** onto **b**, can be expressed as:

$$\text{Proj}_b\mathbf{a} = \left(\frac{\mathbf{a}\cdot\mathbf{b}}{\|\mathbf{b}\|}\right)\left(\frac{\mathbf{b}}{\|\mathbf{b}\|}\right) = \frac{\mathbf{a}\cdot\mathbf{b}}{\mathbf{b}\cdot\mathbf{b}}\mathbf{b}$$

This equation contains three expressions:

- The left expression is the maths notation for vector projections. A mnemonic for remembering this format is to think of "**a** onto **b**" translating to "**a** on top of **b**", which reminds me to place the **b** at the bottom.
- The center expression shows that the magnitude of **b** is used as a denominator to scale two terms: **a** · **b** and **b**. This is important to realize because if *b* is a unit vector (meaning that you are projecting onto a unit vector), then the projection of "**a** onto *b*" simplifies to simply (**a** · *b*) *b*. This two-term form is nice because the left term is a scalar that completely accounts for the magnitude of the resulting vector (including the sign), and the right term is a unit vector that completely accounts for the direction.
- The right expression is nice because it combines the two separate denominator terms into a single denominator dot product. It also manages to group all the scalars together on the left. This is a great way to show that when projecting onto **b**, we are really just trying to calculate how much to scale **b**. All of our computations in the division section simply determine the scaling factor to apply to **b**.

The magnitude of $\text{Proj}_b\mathbf{a}$ is sometimes referred to as the scalar projection. Using SohCahToa (which we will explore in a future chapter) you can express this value as $\cos(\theta) = \|\text{Proj}_b\mathbf{a}\| / \|\mathbf{a}\|$, where θ is the angle between the vector **a** and the vector you are projecting **a** onto. Rewriting the expression produces $\|\text{Proj}_b\mathbf{a}\| = \cos(\theta) * \|\mathbf{a}\|$.

Affine Projections

I want to address a major caveat that caused me to fail an interview some years ago. If projecting onto a structure with position, then that structure needs to contain the origin for the projection to work. This is somewhat complicated, but critical to understand. For example, if you are projecting onto a line, line-segment, or ray, then you need to first translate it to the origin before projecting onto it. Vectors have no position and do not require this step but things like lines, planes, etc. do have position so this step is imperative.

If line AB does not go through the origin, here are the steps to projecting vector **p** onto AB:

1) Subtract A from **p** to obtain **p**'
2) Then project **p**' onto AB for partial solution **h**.
3) Finally, add A to **h** (undoing our offset earlier) for the final answer.

In the first step we needed to subtract a point of AB from **p**, so A was an easy choice. But we also could have used B or any point along AB. This may seem like a very strange step but basically it is just a walk to the origin. For example, if A = (0, 0, 10), then it is 10 units up the z-axis, away from the origin. By subtracting A from **p** we are undoing that 10 unit difference, which then allows us to project onto AB as if AB started at the origin. The final step undoes that translation to convert from our partial solution to a final solution.

You can watch a worked example of this process on my YouTube channel in the video titled "This math problem lost me my job (Affine Projections)" [2].

Perpendicular Projections

In the projection examples we have thus far discussed, the projection of **a** onto **b** resulted in a vector representing the component of **a** that was parallel to **b**. In other words, the result of the projection of **a** onto **b** is always a vector parallel to **b**. This has led to some folks referring to this projection as the "parallel projection". In contrast, you might come across something called a "perpendicular projection" aka "orthogonal projection" in which we actually want the component of **a** that is perpendicular to **b**. In R^3, this is easy to compute. To find the perpendicular projection of **a** onto **b**, simply subtract the parallel projection of **a** onto **b** from **a**. This process removes any part of **a** that was parallel to **b**, leaving only the perpendicular parts.

Projection Inequality

The following projection inequality is another peculiar and extremely useful application of the dot product. Here is the equation:

$$0 \le (AB \cdot AM) \le (AB \cdot AB)$$

FIGURE 7.7   Three points, A, B, and M.

This expression is true when M, projected onto line AB, falls between A and B. For example, with reference to Figure 7.7…

- If M was to the left of A, then (AB · AM) < 0 and the first inequality would not be true.
- If M was to the right of B, then (AB · AM) > (AB · AB) and the second inequality would not be true.

So, this expression is an effective mathematical way to check "if M is between A and B". As we will see further into this book, this method is great for testing if a point is within a polygon. That's because if a point is on the same side of all of a polygon's edges, then it's inside the polygon. And, as we will see, this approach can be generalized into higher dimensions as well. If a point is on the same side of all of a polyhedron's faces, then it's inside the polyhedron.

Let's think about why this works… The dot product can represent a projection. The projection formula normally has a denominator that prevents us from expressing such a simplified inequality. But since we are projecting onto AB in both cases of the dot product, those denominators cancel out in a way that allows us to simplify the expression into the form provided. And when you think about M projected onto AB as M'. All this inequality is asking is if M' is between A and B.

- If M' is to the left of A, then AM' will have a 180° angle from AB causing a negative dot product result and failing the first inequality.
- If M' is to the right of B, then AM' will have a 0° angle from AB, so the dot product will be 1 and we can compare just the magnitudes. Since M' will be further from A than B, the magnitude of (AB · AM') will exceed the magnitude of (AB · AB), failing the second inequality.

You can watch a worked example of this equation on my YouTube channel in the video titled "Applications of the dot product inequality" [3].

3D CROSS PRODUCT

The 3D cross product aka "outer product" or simply "vector product" returns a vector perpendicular to two input vectors, multiplied by the product of their magnitudes, multiplied by the sine of the angle between them. A mnemonic to remember to use the sine function for the <u>cross</u> product, is to think that the 's' in sine looks like two <u>cross</u>ed curves.

The 3D cross product formula: $\mathbf{a} \times \mathbf{b} = \|\mathbf{a}\| \|\mathbf{b}\| \sin(\theta) \, \boldsymbol{n}$

- $\boldsymbol{n}$ is a unit vector perpendicular to both $\mathbf{a}$ and $\mathbf{b}$.
- θ is the angle between $\mathbf{a}$ and $\mathbf{b}$.

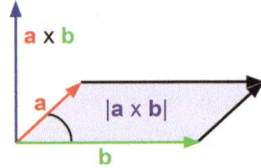

FIGURE 7.8   The area of a parallelogram formed by two vectors, **a** and **b**, is ‖**a** x **b**‖.

Unlike the dot product, which returned a scalar (a number), the cross product returns a vector. In Figure 7.8, the blue vector, **a×b**, is the cross product of vectors **a** and **b**. Observe how it is perpendicular to both of them.

The magnitude of the cross product's resulting vector represents valuable information as well. It's equal to the area of the parallelogram formed by the input vectors, as shown in Figure 7.8. If **a** and **b** are parallel, then the resulting vector will be the zero vector, **0**. This makes sense because two colinear lines would result in a parallelogram of zero height, and therefore zero area. If **a** and **b** are perpendicular, then the parallelogram is a rectangle, so cross product magnitude will be simply "base * height" or ‖**a**‖ * ‖**b**‖.

Practice Problem 7.8: If you take the cross product of two unit vectors, is the resulting vector going to be unit length?

> **Solution**: Another way to phrase this question might be, "In which case would the parallelogram formed by the two unit vectors have an area of 1"? Think of the formula: area = base * height. For the area to be 1, the base and height would both need to be 1 as well. The two unit vectors are of length 1 so the answer is yes… Sike! Did I fool you? Yes, it is true that area = base * height. But two sides of a parallelogram are not necessarily their base and height. That's only true for rectangles. Therefore, the true answer is that the resulting vector will only be unit length if the angle between the input vectors is a right angle such that they form a rectangular area.

You may have noticed that this formula returns the angle between two vectors just like the dot product formula but remember that in this case it is the sine of the angle instead of the cosine. As you will soon see, a cross product takes a lot of computation, about $O(n^2)$ floating point operations (FLOPs) where n is the number of entries in each vector. For comparison, the dot product is only about $O(n)$ FLOPs (much faster). Therefore, if you're just trying to get the magnitude of the angle, it's more efficient to just use the dot product.

However, the cross product can provide us with the direction of theta which the dot product cannot. The directionality of the angle theta can be determined by the directionality of *n*, by using the right-hand rule.

Anti-Commutativity and the Right-Hand Rule

Unlike the dot product, the cross product is anti-commutative. Meaning $\mathbf{a} \times \mathbf{b} = -\mathbf{b} \times \mathbf{a}$. In other words, the order of input vectors matters, and if you switch the order of the vectors, one will have its sign flipped. If you think consider a two non-collinear input vectors, there is one line in R^3 that is perpendicular to both of them. The anti-commutative nature of the cross product means that the order of input vectors will determine if a vector is returned going up that line or down that line. We can determine which vector we will get by using the right-hand rule.

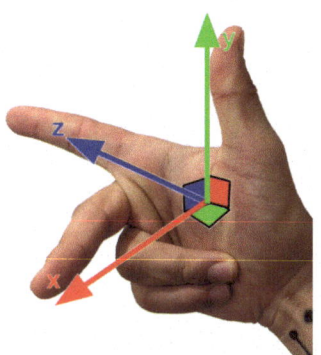

FIGURE 7.9   The right-hand rule demonstrated.

To use the right-hand rule, you require a right hand with an intact thumb, pointer finger, and middle finger. If you don't have a right hand, you can imagine one. First, orient your pointer finger and middle finger toward each of the input vectors. In the example of Figure 7.9, the pointer finger is oriented along the blue line (leftward) and the middle finger is oriented along the red line (toward yourself). With your fingers oriented, your thumb will naturally point in the direction of the cross product result. In the case of the example, this result extends upward.

To demonstrate the anticommutative property, you can now try to switch the order of red and blue. First aim your middle finger along the blue line (leftward) and now this time aim your middle finger along the red line (away from you). Make sure you're aiming it away from you, rather than toward you like last time in order to honor the negative sign in $\mathbf{a} \times \mathbf{b} = -\mathbf{b} \times \mathbf{a}$. You'll find that, while your pointer finger is pointed leftward, in order for your right hand's middle finger to point away from you, you will need to invert your hand upside down and the result is that your thumb will point down instead of up.

The right-hand rule only works in right-handed coordinate systems. In left-handed coordinate systems, you can perform the same steps with your left hand which is called the left-hand rule. Most engines I have worked in such as Unity, Unreal, and the Call of Duty engine are left-handed. But some, like idTech, have been right-handed.

Practice Problem 7.9: How would you tell if a game engine uses a right-handed coordinate system?

> **Solution**: Using the right-hand rule, I would aim my right-handed pointer finger in the direction of the x-axis, then my middle finger in the direction of the y-axis. If my thumb then extended in the positive direction of the z-axis, I would know the system is right-handed. If instead my thumb was pointed in the negative direction of the z-axis, I would know the system is left-handed.

Additively Distributive

Like the dot product, the cross product is additively distributive $(\mathbf{a} \times (\mathbf{b} + \mathbf{c})) = (\mathbf{a} \times \mathbf{b}) + (\mathbf{a} \times \mathbf{c})$.

Remember that the cross product's result vector is perpendicular to the two input vectors. So here we are saying that the cross product of two vectors is going to be the same as taking the sum of cross products between one of those vectors crossed with the components of the other vector.

Triple Product

Do you remember how a cross product's magnitude is equal to the area of the parallelogram formed by the two input vectors? I want to show you something like an abstraction of that concept in 3D. There's this really cool product called the triple product. It's formed by using the dot product AND the cross product; crazy, I know! It looks like this: $\mathbf{a} \cdot (\mathbf{b} \times \mathbf{c})$. And this expression evaluates to the signed volume of the parallelepiped formed by $\mathbf{a}$, $\mathbf{b}$, and $\mathbf{c}$. The below image is an example of such a volume.

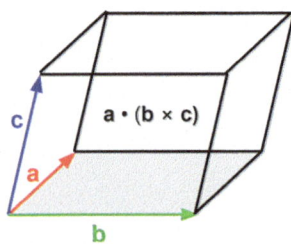

FIGURE 7.10 The triple product.

Cross Product Computation

So far, we have discussed at length what the components of a cross product represent and what its result equals, but not how to actually compute it. In order compute a cross product, we need to use the matrix determinant formula. We'll cover the meaning of a matrix determinant in the matrix chapter. For now, understand that the determinant of a 2x2 matrix can be expressed as a relationship of diagonally adjacent values.

$$\mathbf{M} = \begin{bmatrix} a & b \\ c & d \end{bmatrix}$$

With **M** defined above, $\det(\mathbf{M}) = ad - bc$.

For a 3x3 matrix, the determinant formula is a bit more complicated, but it actually simplifies down to an expression of several 2x2 determinants. To compute a cross product result, we define a 3x3 matrix like this:

$$\mathbf{a} \times \mathbf{b} = \begin{vmatrix} i & j & k \\ a_1 & a_2 & a_3 \\ b_1 & b_2 & b_3 \end{vmatrix} = \begin{vmatrix} a_2 & a_3 \\ b_2 & b_3 \end{vmatrix} i + \begin{vmatrix} a_1 & a_3 \\ b_1 & b_3 \end{vmatrix} j + \begin{vmatrix} a_1 & a_2 \\ b_1 & b_2 \end{vmatrix} k$$

In this equation, the unit vector variables i, j, and k represent the three components of the 3D vector result. When we solve the determinant, it will simplify to the form: $\mathbf{a} \times \mathbf{b} = ei + fj + gk$. Where i, j, and k are unit vectors and e, f, and g are scalar factors as shown below:

$$(a_2b_3 - a_3b_2)i - (a_1b_3 - a_3b_1)j + (a_1b_2 - a_2b_1)k$$

This process is all done by code, but it's important to underscore that there is a lot of computation here. Cross products are very useful but also very expensive.

Practice Problem 7.10: If given a player's forward vector **f** and the vector from a player to a target, **t**, explain how you would derive the signed magnitude of the angle the player must turn to face the target.

Hint: Basically, you need to determine how much we need to rotate **f** so that it is collinear with **t**.

Solution: The first part of the problem is to get the angle between **f** and **t** which we can do by using the dot product. That angle is unsigned, so we now just need to determine if the rotation is clockwise or counter-clockwise. This second part requires two steps. First we determine the plane we are rotating within. This plane will be defined by the normal vector perpendicular to both **f** and **t**. To find that vector we can use the cross product:

```
Vector3 planeNormal = Vector3.Cross(f, t);
```

Now to see if our rotation is clockwise from the player's perspective we just need to test if the plane normal is in the same direction as our character's up direction.

bool shouldTurnClockwise = $\text{Vector3.Dot}(\text{planeNormal}, \text{character.up}) > 0;$

The problem statement did not indicate that the character had an up direction. Frequently questions of this nature will not include details like this and interviewers will instead expect candidates to either ask questions to clarify those details or make assumptions based on their experience.

This shouldTurnClockwise derivation works because the dot product returns a positive value only when the input vectors are separated by an acute angle. And if the rotation plane's normal and our player's up are separated by an acute angle, then that means they point in the same general direction therefore we are rotating clockwise.

Our final answer is that we need to rotate around the plane defined by planeNormal for angle theta in the direction dictated by shouldTurnClockwise.

2D CROSS PRODUCT

The 2D cross product, occasionally called the "perpendicular product" or "wedge product," returns a scalar that represents the magnitude of a vector perpendicular to two input vectors in 2D space.

In the previous section we learned about the 3D cross product, and you might think that the 2D cross product is the same thing, just in 2D. But you'd be wrong! It is true that the 2D cross product takes 2D vectors instead of 3D vectors, but this product returns a scalar instead of a vector. Despite this notable difference, the 2D cross product has a similar role to the 3D cross product. The 2D cross product's result will tell us if its first input vector is on the left or right side of its second input vector, just like the 3D cross product can be used to determine the direction of an angle between input vectors.

2D Cross product formulae:

$$\mathbf{a} \times \mathbf{b} = (\mathbf{a}.x * \mathbf{b}.y) - (\mathbf{b}.x * \mathbf{a}.y)$$
$$AB \times CD = (B.x - A.x)(B.y - A.y) - (D.x - C.x)(D.y - C.y)$$

The first formula is for vectors, the second formula is for line segments.

The Assuming a left-handed coordinate system the magnitude of **a**×**b** is...

- Positive if the **a** is on the left-hand side of **b**.
- Negative if **a** is on the right-hand side of **b**.
- Zero if **a** and **b** are collinear.

Practice Problem 7.11: You are given two points: A and B, and a line segment defined by a start and end point: S and E. Determine if A and B are on the same side of line segment SE.

Solution: We can use the 2D cross product to determine what side of a line a point is on. First, we test point A. AS × AE = a. If a < 0, then the point lies on one side of the line, and if a > 0, then it lies on the other side. If a=0, then the point lies exactly on the line. Now we can repeat this step for the other point: B. BS × BE = b. Now we can compare the results, a and b, to see if they are the same polarity (aka both negative or both positive) which will tell us if A and B are on the same side of SE.

Practice Problem 7.12: You are given a triangle formed by points ABC. And a separate point, P. Determine if P lies within the triangle.

Solution: In order to test which side of AB the point P is on, we can use the 2D cross product to compute AP × AB. If P is on the same side of all the triangle's edges, then it is in the triangle. Therefore, our solution is to just repeat this check for the remaining sides. This approach actually works for all polygons! It just requires you to go through the points in order. Remember the order matters a lot because if you compare a point to AB instead of BA those two line segments are literally going in opposite directions! So, make sure to pick a consistent order (like clockwise) to test the points and stick to it, such as comparing P against AB, BC, and finally CA.

This is a very common problem, and alternative approaches have various benefits/drawbacks. We will discuss an alternative Barycentric coordinate approach in a future chapter on polygons.

PRACTICE PROBLEMS

Vector maths is the most likely topic to come up in a gameplay engineering interview. To help you further prepare, I have included a gauntlet of questions in this section. Some of these questions take longer to solve, so they are less likely to come up during an onsite in-person interview, but they may be asked of you in a take-home test. The important part to solving each of these problems is devising a viable approach, rather than implementing the exact minutia of a solution.

Practice Question 7.13: Given vectors **a** and **b**, determine if they are ∥ or ⊥. If you have not seen these symbols before, they mean parallel and perpendicular, respectively.

Solution:

$\mathbf{a} \cdot \mathbf{b} = 0$ if ⊥.

$\mathbf{a} \times \mathbf{b} = \mathbf{0}$ if ∥ (this is the same as checking if $\|\mathbf{a} \times \mathbf{b}\| = 0$).

Practice Problem 7.14: Given line AB, find any other line that is perpendicular to AB.

Hint: If two vectors are perpendicular, how can we express that in an equation to solve for the second vector?

Solution: To keep things simple and call the vector that defines line AB, aka B − A, **v**. Now let's handle the edge cases. If all components of **v** are zero, then **v** is the zero vector and we can return a sentinel value, like [0,0,1]. If one or two of **v**'s components are zero, then we can simply return a vector with corresponding components that are non-zero for the input vector's zero values and non-zero for the input vector's zero values. For example, [0, 1, 2] can be given [1, 0, 0] and [0, 0, 1] can be given [1, 1, 0].

Finally, we are left to deal with the case when **v** has no zero components. To find a vector perpendicular to **v**, $\mathbf{v}_\perp$, we need to solve for a dot product of zero. Here's the formula:

$$\left(\mathbf{v}.x * \mathbf{v}_\perp.x\right) + \left(\mathbf{v}.y * \mathbf{v}_\perp.y\right) + \left(\mathbf{v}.z * \mathbf{v}_\perp.z\right) = 0$$

Now let's isolate the x component we need.

$$\mathbf{v}_\perp.x = \left(0 - \left(\mathbf{v}.y * \mathbf{v}_\perp.y\right) - \left(\mathbf{v}.z * \mathbf{v}_\perp.z\right)\right) / \mathbf{v}.x$$

To satisfy the equation, choose any random non-zero values for $v_\perp.y$ and $v_\perp.z$. Then plug them in to solve for $v_\perp.x$.

$a \times b = 0$ if $\parallel$ (this is the same as checking if $\|a \times b\| = 0$).

Practice Problem 7.15: Find the distance between a line segment, BC, and point, A.

Solution: This is a classic question, and one you need to feel comfortable solving. I absolutely recommend drawing a diagram for questions like these.

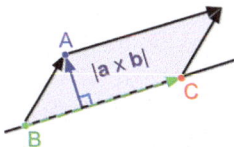

FIGURE 7.11 A parallelogram with base BC.

The first approach I would like to show you uses the parallelogram area formula. Since we know that the magnitude of the cross product is equal to the area of the parallelogram defined by the input vectors, we can determine that area with |BA x BC|. Then since we know that the area of a parallelogram is also its base * height, we can use this area with the base length, |BC| to derive the height length. And the cool part is that our height length is actually the distance to A. We know this because the shortest distance between a point and a line is always going to be along a line perpendicular to our starting line. And that also happens to be the height of our parallelogram. Here are the steps (see Figure 7.11):

Area = Base * Height

$\|BA \times BC\| = \| BC \| *$ Height

Height $= \|BA \times BC\| / \| BC \|$

You can view a worked solution of this approach in the video "Cross Product and Distance from a Point to a Line" by turksvids on YouTube [4].

An alternative approach to this problem is to use vector projection. If we project BA onto BC then we will get vector BD, where D is the point on line BC that is closest to A. With BD known, can then trivially compute D, because D = B + BD. And all that remains is to check the distance between D and A. This approach is far more efficient than the previous approach because we avoid the expensive cross product. Here are the steps (see Figure 7.12):

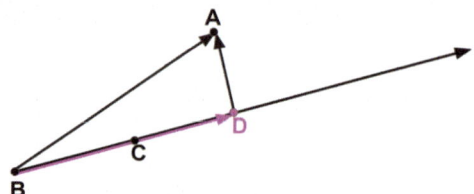

FIGURE 7.12   The projection of A onto BC is D.

Because we are projecting onto a line, line segment, or ray (and not a vector) we need to use the affine projection steps in case BC does not intercept the origin.

First let's define vector **p** as A-B and project it onto BC:

1) Subtract B from **p** to obtain **p'** → **p'** = A-B-B
2) Then project **p'** onto BC for partial solution **h** → **h** = proj$_{BC}$A-B-B
3) Finally, add B to **h** (undoing our offset earlier) for the final answer. → BD = B + proj$_{BC}$A-B-B

Now that we have BD, we can calculate D = B + BD. And finally, we can calculate the distance between D and A, which is the same as the magnitude of DA, ‖DA‖.

WORKS CITED

[1] P. Danilyuk, "A Robotic Arm and a Chessboard," *Pexels*, Aug. 2023. [Online]. Available: https://www.pexels.com/photo/a-robotic-arm-and-a-chessboard-8438868. [Accessed: Aug. 15, 2024].

[2] M. Ventures, "This math problem lost me my job (Affine Projections)," *YouTube*, Aug. 16, 2024. [Online]. Available: https://www.youtube.com/watch?v=eAmwF4oc3DA&ab_channel=MatthewVentures

[3] M. Ventures, "Applications of the dot product inequality," *YouTube*, Aug. 16, 2024. [Online]. Available: https://youtu.be/srqhPL4nWlI?si=BKCus1YN8SCShNBY

[4] T. A. Turksvids, "Cross Product and Distance from a Point to a Line," *YouTube*, Dec. 13, 2023. [Online]. Available: https://www.youtube.com/watch?v=tYUtWYGUqgw&ab_channel=turksvids. [Accessed: 16 Aug. 2024].

Maths
Matrices

8

Matrices are multidimensional number structures. They are extremely important to game development, and you need to be very good friends with them (if not lovers).

MATRIX MULTIPLICATION

- It is associative aka $(\mathbf{AB})\mathbf{C} = \mathbf{A}(\mathbf{BC})$
- It is not commutative $\mathbf{AB} \mathrel{!=} \mathbf{BA}$

When we refer to matrices, we often reference their bounds as m×n, which stands for the "row count by column count" aka "height by width".

If you take a matrix **A** that is m×n then it can only be multiplied against a matrix **B** if that matrix has n rows. In other words, if **A**'s dimensions are e×f and B's are m×n then when we put them together like this (e×f)(m×n), f and m (the two numbers where the dimensions meet) need to be equal. And we also know that the resulting matrix will be of dimension e×n. So, if I told you we have a 3×1 and a 1×7, you immediately know that multiplying them is valid (because the 1's are equal) and that the result will be 3×7 (because those are the values on either end).

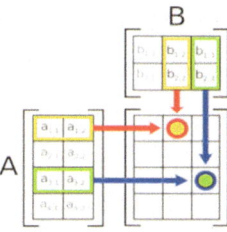

FIGURE 8.1 Matrix multiplication forms cells via the dot product of a row and column from each operand.

DOI: 10.1201/9781003565550-10

To perform matrix multiplication, you calculate the resulting matrix's cells one at a time. To perform **AB** (as shown in Figure 8.1), you first determine the row of **A** and the column of **B** that correspond to the current result cell. Then go left-to-right through each entry of **A**'s row and top-to-bottom through each entry of **B**'s column to create a series of pairs. Multiply the values in each pair and then sum the results to obtain the result cell's value. This process of multiplying and summing the pairs is equivalent to calculating the dot product of the selected row and column vector.

DETERMINANT

The determinant of a matrix is a scalar of the volume/area represented by a matrix. If it's non-zero, then the matrix has an inverse. If it's positive, then the transformation will maintain the orientation or "handedness" of the input. And if it's negative, the transformation will reverse the handedness. The main role of the determinant is for determining a matrix's inverse. A determinant is denoted by vertical bars around a matrix such as |**A**|, though sometimes you might see it written as det(**A**). In the previous chapter in the section "Cross Product Computation" we covered a specific matrix determinant calculation used to compute the cross product, but here are the general formulae for 2×2 and 3×3:

$$\mathbf{M} = \begin{bmatrix} a & b \\ c & d \end{bmatrix} \qquad |\mathbf{M}| = ad - bc$$

$$\begin{vmatrix} a & b & c \\ d & e & f \\ g & h & i \end{vmatrix} = a\begin{vmatrix} e & f \\ h & i \end{vmatrix} + b\begin{vmatrix} d & f \\ g & i \end{vmatrix} + c\begin{vmatrix} d & e \\ g & h \end{vmatrix}$$

ORTHOGONAL MATRICES

All rotation matrices are orthogonal aka orthonormal. This means they are real square matrices whose columns and rows are orthonormal vectors. Two vectors are orthogonal if their dot product is 0. Two vectors are orthonormal if they are orthogonal, and unit length.

Orthogonal matrices have the following properties:

- Each column/row vector is unit length. Therefore, if you take anyone of the vectors **v**, then $\mathbf{v} \cdot \mathbf{v} = 1$
- Each column/row vector is orthogonal to all other columns/rows. Therefore, if you take one row vector **v**, and any other row vector **y**, then $\mathbf{v} \cdot \mathbf{y} = 0$. Furthermore, the collection of orthogonal columns/rows forms a basis.

- They are square (their row count equals their column count).
- Their inverse is their transpose aka $M^T = M^{-1}$ which is a huge deal! We are going to go into more details about matrix inverses later, but for now, just know that the inverse of a matrix is expensive to calculate. So, being able to calculate the transpose instead means major savings.
- Their determinant is either +1 (preserving orientation) or -1 (indicating a reflection). This implies that a vector's scale is unchanged by a rotation, but that its direction may be changed. This intuitively makes sense (things don't become smaller when you turn them, but they may change direction).
- The product of an orthogonal matrix M with its transpose will yield the identity matrix aka $M \bullet M^T = I$.

MATRICES INVERSES

Here are two formulae to calculate a matrix inverse. One is for 2×2 matrices and one is for an arbitrary block upper triangular matrix. "Upper triangular" in this 2×2 case just means that the bottom left value is zero.

$$A = \begin{bmatrix} a & b \\ c & d \end{bmatrix}, \; A^{-1} = \frac{1}{|A|} \begin{bmatrix} d & -b \\ -c & a \end{bmatrix}$$

$$M = \begin{bmatrix} M_{11} & M_{12} \\ 0 & M_{22} \end{bmatrix}, \; M^{-1} = \begin{bmatrix} M_{11}^{-1} & M_{11}^{-1} \times M_{12}^{-1} \times M_{22}^{-1} \\ 0 & M_{22}^{-1} \end{bmatrix}$$

We'll return to this upper triangular concept when discussing rigid body transform matrices. For now, the important point to understand is that if we can get a matrix into a block upper triangular format, it can allow us to use upper triangular block formula, which can save considerable time. Without this formula we would need to use the generic 3×3 inverse formula shown below. It looks gargantuan, because it is. The generic 3×3 inverse formula includes a staggering nine separate determinant calculations, therefore avoiding it is very valuable.

$$A = \begin{bmatrix} a_{11} & a_{12} & a_{13} \\ a_{21} & a_{22} & a_{23} \\ a_{31} & a_{32} & a_{33} \end{bmatrix}, \; A^{-1} = \frac{1}{|A|} \begin{bmatrix} \begin{vmatrix} a_{22} & a_{23} \\ a_{32} & a_{33} \end{vmatrix} & \begin{vmatrix} a_{13} & a_{12} \\ a_{33} & a_{32} \end{vmatrix} & \begin{vmatrix} a_{12} & a_{13} \\ a_{22} & a_{23} \end{vmatrix} \\ \begin{vmatrix} a_{23} & a_{21} \\ a_{33} & a_{31} \end{vmatrix} & \begin{vmatrix} a_{11} & a_{13} \\ a_{31} & a_{33} \end{vmatrix} & \begin{vmatrix} a_{13} & a_{11} \\ a_{23} & a_{21} \end{vmatrix} \\ \begin{vmatrix} a_{21} & a_{22} \\ a_{31} & a_{32} \end{vmatrix} & \begin{vmatrix} a_{12} & a_{11} \\ a_{32} & a_{31} \end{vmatrix} & \begin{vmatrix} a_{11} & a_{12} \\ a_{21} & a_{22} \end{vmatrix} \end{bmatrix}$$

Additional matrix inverse facts:

- If $\mathbf{M} = \mathbf{R} \times \mathbf{T}$ then $\mathbf{M}^{-1} = \mathbf{T}^{-1} \times \mathbf{R}^{-1}$, meaning that taking the inverse reverses the order of the matrix operands.

ROTATION MATRICES

Rotation matrices are commonly used to store rotational data in 3D space. They offer a straightforward way to apply and understand rotations, especially when transforming an object's orientation between different coordinate systems (like world space and local space).

However, rotation matrices are not always ideal for dynamic interpolation of rotations during runtime. This is because Euler (pronounced "Oiler") angles, which are often used in conjunction with rotation matrices, are prone to a problem known as gimbal lock. Gimbal lock occurs when two of the three rotational axes align, causing a loss of one degree of freedom. This alignment effectively restricts rotational movement, requiring additional adjustments to regain the lost freedom.

Rotation matrices are orthogonal and furthermore always have a determinant of 1. Remember that since they are orthogonal, they are skew-symmetric, meaning that their transpose is also their inverse.

On previous interviews, I have been asked to construct rotation matrices from scratch, so this is a format with which you should probably familiarize yourself. I recommend checking out the Khan Academy lesson titled "Linear Transformation Examples: Rotations in R2." By Sal Khan and the YouTube video titled "What are affine transformations?" by Leios Labs [2], [3]. We'll cover the common cases of 2D and 3D rotation matrix construction. Below is the 2D case which represents a counterclockwise rotation through an angle "θ" about the origin.

$$\mathbf{Rv} = \begin{bmatrix} \cos\theta & -\sin\theta \\ \sin\theta & \cos\theta \end{bmatrix} \begin{bmatrix} x \\ y \end{bmatrix} = \begin{bmatrix} x\cos\theta - y\sin\theta \\ x\sin\theta + y\cos\theta \end{bmatrix}$$

Things get a little trickier in 3D where you will need to use a separate formula to represent each of the axes of rotation (shown below). Something that helped me conceptualize abstract rotations from 2D to 3D is realizing that every rotation is always going to be through a plane around an n-2 dimensional subspace. That is true whether you are in 2D (n=2) or 4D (n=4) or even some imaginary 99D. Here's some examples:

- Rotations in 2D (n=2) are always through a plane around a point. It's a point because if n=2, then n-2=0 and a 0-dimensional subspace is a point.
- Rotations in 3D (n=3) are always through a plane around a line. It's a line because if n=3, then n-2=1 and a 1-dimensional subspace is a line.

So, when you look at the 3D rotation matrices below, consider how each is designed to define which line we are rotating around: $\mathbf{R}_x$ is for the x axis, and so on.

$$\mathbf{R}_x(\theta) = \begin{bmatrix} 1 & 0 & 0 \\ 0 & \cos\theta & -\sin\theta \\ 0 & \sin\theta & \cos\theta \end{bmatrix}$$

$$\mathbf{R}_y(\theta) = \begin{bmatrix} \cos\theta & 0 & \sin\theta \\ 0 & 1 & 0 \\ -\sin\theta & 0 & \cos\theta \end{bmatrix}$$

$$\mathbf{R}_z(\theta) = \begin{bmatrix} \cos\theta & -\sin\theta & 0 \\ \sin\theta & \cos\theta & 0 \\ 0 & 0 & 1 \end{bmatrix}$$

Once we understand these axis-specific rotation matrices we can combine them to represent any complex rotation. Let's take a moment to analyze how these matrices are constructed so that you will be able to build one if requested in an interview.

Notice that a specific row and column of each matrix corresponds to the axis it represents. For example, in $\mathbf{R}_x$ all of the first row and first column values are '0', except at the element where the row and column intersect. That's because it is the first row and first column which correspond to the x-axis. Observe similar empty columns/rows within the other two matrices. If you can remember to set the values of that row and column to '0' (except for the '1' where they intercept) you will know five of the nine values within a rotation matrix.

The remaining four values are always in a box formation with cos on the top left and bottom right, and sin in the top right and bottom left. The final thing to remember is that the top left sin is negative for $\mathbf{R}_x$ and $\mathbf{R}_z$, while the bottom left sin is negative for $\mathbf{R}_y$. I developed the mnemonic, "far right groups have negative signs because they never ask 'why?'" to help me recall that the top-right sin is negative unless it is a y-axis rotation, in which case it's the bottom left sign that is negative instead.

Practice Problem 8.1: Construct a rotation matrix that represents a rotation of 90° counterclockwise through the xy-plane.

Solution:

$$\mathbf{R}_z(90°) = \begin{bmatrix} \cos 90° & -\sin 90° & 0 \\ \sin 90° & \cos 90° & 0 \\ 0 & 0 & 1 \end{bmatrix}$$

You can multiply rotation matrices together to create a rotation matrix that is the result of both rotations. As you may recall, the order of this operation matters because unlike scalar multiplication, matrix multiplication is not commutative $\mathbf{AB} \mathrel{!=} \mathbf{BA}$. The input vector you want to rotate will always go on the right-hand side of a matrix like the one shown in the earlier 2D case of $\mathbf{Rv}$.

Practice Problem 8.2: Construct a rotation matrix that represents a rotation of 45° counterclockwise through the yz-plane, followed by a rotation of 90° counterclockwise through the xy-plane.

Solution: Remember, the first rotation to applied is placed on the right-hand side.

$$\mathbf{R}_z\left(90°\right)\mathbf{R}_x\left(45°\right) = \begin{bmatrix} \cos 90° & -\sin 90° & 0 \\ \sin 90° & \cos 90° & 0 \\ 0 & 0 & 1 \end{bmatrix}\begin{bmatrix} 1 & 0 & 0 \\ 0 & \cos 45° & -\sin 45° \\ 0 & \sin 45° & \cos 45° \end{bmatrix}$$

$$\mathbf{R}_z\left(90°\right)\mathbf{R}_x\left(45°\right) = \begin{bmatrix} \cos 90° & -\sin 90° * \cos 45° & \sin 90° * \sin 45° \\ \sin 90° & \cos 90° * \cos 45° & -\cos 90° * \sin 45° \\ 0 & \sin 45° & \cos 45° \end{bmatrix}$$

$$\mathbf{R}_z\left(90°\right)\mathbf{R}_x\left(45°\right) = \begin{bmatrix} 0 & -\sin 90° * \cos 45° & \sin 90° * \sin 45° \\ 1 & 0 & 0 \\ 0 & \sin 45° & \cos 45° \end{bmatrix}$$

Something really cool about rotation matrices is that each column of the matrix represents an axis vector and together they define a basis. This is rather important, so I have illustrated the concept below. On the left we have the standard basis and on the right, we the standard basis after a 90° counterclockwise rotation around the z-axis. As you can see, the vectors representing each of these columns illustrate the rotation. The red vector has moved to the right side of the image, where the green vector was. I recommend finding an online plotter and taking some time to explore how rotation matrices can affect vectors by playing with values and plugging them in.

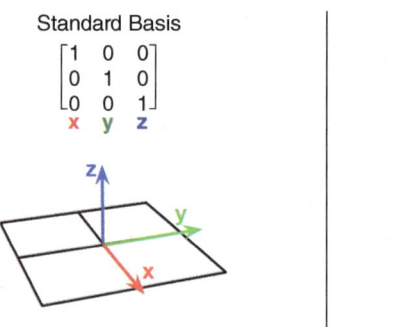

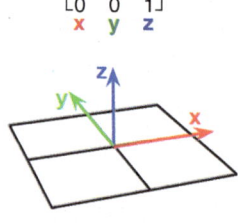

Practice Problem 8.3: Given a local forward vector, **f**, and the world-space up, **u**, create a coordinate basis for the local transform.

Solution: First, we already have **f** which is one column of the 3×3. To get the local right (or left vector) we can perform the cross product of **f** and **u**, let's call it **r**. Finally, to get the local up (or down) vector we can perform the cross product of **f** and **r**, let's call it $\mathbf{u}_{local}$. The three local vectors, **f**, **r**, and $\mathbf{u}_{local}$, can now be arranged as columns in a matrix, forming a local coordinate basis.

You may have noticed some parenthetical ambiguity in that solution. That's because we can't assume that the system is right-handed. In a right-handed system to interpret **f×u**, you aim your pointer and middle fingers in the direction of the first and second operands respectively. As you twist your hand in this process, so that your pointer finger aims forward and your middle finger rotates clockwise to aim upward, you will observe your palm opening to face upward as well. The result is your thumb will be oriented right. This means that **f×u** returns the right vector in a right-handed system, and implies that it would instead return the left vector in a left-handed system.

TRANSLATION MATRICES

$$\begin{bmatrix} 1 & 0 & t_x \\ 0 & 1 & t_y \\ 0 & 0 & 1 \end{bmatrix} \begin{bmatrix} a_x \\ a_y \\ 1 \end{bmatrix} = \begin{bmatrix} a_x + t_y \\ a_y + t_y \\ 1 \end{bmatrix}$$

A translation matrix, when multiplied with a position vector, will translate the position. In the above example, the left-hand-side translation matrix translates vector **a** by vector **t**, which geometrically would be expressed as **a** + **t**. This is an example of a 2D translation matrix, but for higher dimensions the design is basically the same. Constructing a translation matrix is simple, place the translation vector on the right-most column. Then fill in the diagonals with 1's and the rest of the matrix entries with 0's in order to maintain the original scale and orientation.

Practice Problem 8.4: Construct a 4×4 3D translation matrix for a translation of [6, 7, 8].

Solution:

$$\begin{bmatrix} 1 & 0 & 0 & 6 \\ 0 & 1 & 0 & 7 \\ 0 & 0 & 1 & 8 \\ 0 & 0 & 0 & 1 \end{bmatrix}$$

SCALING MATRICES

$$\begin{bmatrix} s_x & 0 & 0 \\ 0 & s_y & 0 \\ 0 & 0 & 1 \end{bmatrix} \begin{bmatrix} a_x \\ a_y \\ 1 \end{bmatrix} = \begin{bmatrix} s_x \times a_x \\ s_y \times a_y \\ 1 \end{bmatrix}$$

A scaling matrix, when multiplied with a position vector, will scale the position. In the above example, the left-hand-side transformation matrix scales vector **a** by vector **s**. Usually, we are scaling with equal factors for each component. If that is the case and Sx and Sy are both the same value, w for example, then we can express the scaling as w**a**. This is a 2D example, but for higher dimensions the design is basically the same. You place the scaling vector along the diagonal. You fill the right-hand-side column with 0's (to avoid translating) and the rest of the matrix entries are 0's (to prevent rotating). The bottom right entry, in this case, is not used so we just put a 1 there.

Practice Problem 8.5: Construct a 4×4 3D scaling matrix that represents a scaling of [4, 5, 6].

Solution:

$$\begin{bmatrix} 4 & 0 & 0 & 0 \\ 0 & 5 & 0 & 0 \\ 0 & 0 & 6 & 0 \\ 0 & 0 & 0 & 1 \end{bmatrix}$$

RIGID BODY TRANSFORM MATRICES

The rigid body transform matrix (RBTM) is the most iconic matrix in all of game development. I would advise you to memorize its structure as shown in Figure 8.2. As you will notice, it's designed to be a little larger than a typical 3×3 transformation matrix so that translation data can be easily differentiated from scaling and rotational data.

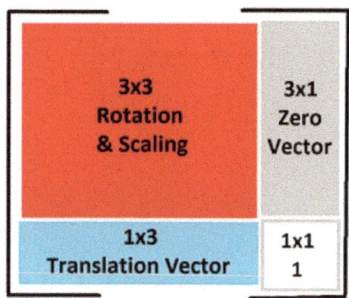

FIGURE 8.2   The rigid body transform matrix.

Let's take a closer look at each part of this block matrix:

- Observe that the top left (red) 3×3 matrix contains both rotation and scaling data. This 3×3 is not quite a rotation matrix because, due to the scaling, its columns are not unit length.
- The translation data is stored separately in the bottom (blue) 1×3 row.
- The remainder elements of the 4×4 transform matrix are filled by a bottom-right (white) value of 1 and a top-right (grey) 3×1 zero vector.

We will soon discuss a few reasons why we often store and manipulate transform data in this format, but first I want to clarify that since the grey & white right hand column's values are static (they will always be a zero vector and a 1) we usually don't store that column at all. This 4×4 matrix is usually "dehydrated" and stored as a 4×3 (for a 25% space savings). It is only rehydrated into its full, glorious, 4×4 form at runtime. For this reason, some developers refer to it as a "4×3" but conceptually, and at runtime, it's really a 4×4.

It's also important to know that depending on the convention of the engine, the (blue) translation vector and (grey) zero vector may be swapped. Later we will look at an example in this alternative format.

There are a number of huge benefits to the RBTM format. One of which is that this block matrix format allows us to easily consider the matrix in parts. When separated into these parts, we can distribute the inverse function in a way that allows us to calculate the total matrix's inverse without ever using that gargantuan matrix inverse formula we discussed previously. The amount of time we save with this approach is wonderous. Let's take a walk through that process now.

First, we split the full RBTM matrix, which we call **M**, into **T** and **R**, matrices which hold **M**'s translation and rotation data respectively.

$$\mathbf{M} = \mathbf{TR}$$

$$\mathbf{M} = \begin{bmatrix} 1 & 0 & 0 & x \\ 0 & 1 & 0 & y \\ 0 & 0 & 1 & z \\ 0 & 0 & 0 & 1 \end{bmatrix} \begin{bmatrix} | & | & | & 0 \\ \mathbf{f} & \mathbf{r} & \mathbf{u} & 0 \\ | & | & | & 0 \\ 0 & 0 & 0 & 1 \end{bmatrix}$$

$$\mathbf{M} = \begin{bmatrix} \mathbf{I} & \mathbf{t} \\ \mathbf{0}^t & 1 \end{bmatrix} \begin{bmatrix} \mathbf{R}_{3\times 3} & \mathbf{0} \\ \mathbf{0}^t & 1 \end{bmatrix}$$

Observe that that the rotation matrix is mainly composed of three column vectors **f**, **r**, and **u**. They are then squeezed into an upper-left 3×3 rotation matrix, within a larger 4×4, when represented in block form. With **M** now split, we can invert it. Please remember that when we distribute an inverse, it will switch the order of terms, so we will see $(\mathbf{TR})^{-1}$ distributed to $\mathbf{R}^{-1}\mathbf{T}^{-1}$.

$$\mathbf{M}^{-1} = \mathbf{R}^{-1}\mathbf{T}^{-1}$$

$$\mathbf{M}^{-1} = \begin{bmatrix} \mathbf{R}_{3\times 3} & \mathbf{0} \\ \mathbf{0}^t & 1 \end{bmatrix}^{-1} \begin{bmatrix} \mathbf{I} & \mathbf{t} \\ \mathbf{0}^t & 1 \end{bmatrix}^{-1}$$

$$\mathbf{M}^{-1} = \begin{bmatrix} \mathbf{R}_{3\times 3}^{\mathrm{T}} & \mathbf{0} \\ \mathbf{0}^t & 1 \end{bmatrix} \begin{bmatrix} \mathbf{I} & -\mathbf{t} \\ \mathbf{0}^t & 1 \end{bmatrix}$$

$$\mathbf{M}^{-1} = \begin{bmatrix} \mathbf{R}_{3\times 3}^{\mathrm{T}} & -\mathbf{R}_{3\times 3}^{\mathrm{T}}\mathbf{t} \\ \mathbf{0}^t & 1 \end{bmatrix}$$

Wow! We managed to express the matrix inverse with just a bit of multiplication and a few transposes. This is really fantastic and saves a huge amount of time compared to solving the generic matrix inverse equation. You can view a worked derivation of this approach in the YouTube video "Math for Game Developers - Faster Matrix Inversions" by Jorge Rodriguez [4].

While the RBTM form has many benefits, there are a few drawbacks. One glaring drawback is that it mangles the rotation and scaling data into a single 3×3 matrix. It can be difficult or impossible to separate that data once it has been mangled depending on what are acceptable rotation and scaling values. That's because though we know each column of a rotation matrix will have a unit length vector, we do not know if that vector will point one direction or another. If scaling is only ever positive, then this is not an issue. However, if we store negative scaling data within a transform matrix then it is indistinguishable from the rotation data and there is no way to differentiate the direction of the rotational basis vectors from the sign of the scaling factor. Think of it like this... sometimes you can't tell if an image was flipped (negatively scale) or rotated. For example, an image flipped horizontally and vertically is identical to the image rotated by 180 degrees. Usually in videogames we don't need to include scaling within the transform data at all. For example, when I was working on Call of Duty, the soldiers were never going to scale in size, so their transform data did not need to hold scaling data. But this design of RBTM was a major problem I ran into while working on Fortnite which led to objects rotating whenever they were scaled negatively. To fix this issue, I ultimately had to save the scale vector into a separate data structure so that the red 3×3 only held rotation data. As with many techniques, RBTM is a tradeoff.

Practice Problem 8.6: Take the generic 3×3 transformation matrix, **M**, below, and deconstruct it into its component scale, rotation, and translation matrices aka its "**TRS**" form. Notice that in this example, the constants are across the bottom row instead of on the far-right column as in our earlier examples. Also notice that it is a 3×3 transformation matrix and not the 4×4 we used previously, this means it is a 2D example (no z coordinates). You may assume that there are only positive scalars.

$$\mathbf{M} = \begin{bmatrix} -2 & -1 & 2 \\ -2 & 1 & -1 \\ 0 & 0 & 1 \end{bmatrix}$$

Hint #1: **M** contains the **TRS** data inside, we need to pull it apart. You're going to want to extract the data in order of the translation matrix, scaling matrix, and rotation matrix.

Hint #2: Two entries of the provided matrix correspond exactly to the x and y translation. Identify those two entries and use them to make a translation matrix.

Hint #3: The upper-left 2×2 of this matrix contains both scaling and rotational data. If you know that a matrix is orthonormal (which all rotation matrices are) then you also know each of its vectors should be of length 1. What can you do to extract the scaling factors out of these vector columns to both determine the scaling that was applied and to restore the original rotation matrix?

Solution: First we can plainly read the translation data from the right most column: [2, -1, 1] means the xy scaling is [2, -1]. We can plug this into the translation matrix format to form our translation matrix, **T**.

For the scaling matrix, calculate the magnitude of the left column to determine the x scaling factor. Repeat for the center column to find the y scaling factor. Plug these two values ($\sqrt{8}$ and $\sqrt{2}$) into the scaling matrix format to form our scaling matrix, **S**. Note that we do not calculate a scaling factor for the right column, because that column contains the translation data only, and no scaling data.

Keep track of those x and y scaling factors. Use them to normalize the left and center columns of **M**. The resulting vectors are used to form our 2×2 rotation matrix. For example, M's x column is [2, -2], since we determined that the x scaling factor is $\sqrt{8}$, R's x column will be the same vector but with each value divided by $\sqrt{8}$, hence [2/$\sqrt{8}$, -2/$\sqrt{8}$]. When dealing with transform matrices, the rotation matrix (which is really just a 2×2) can be extended into a 3×3 which is what you'll see in the solution as matrix **R**.

$$\mathbf{T} = \begin{bmatrix} 1 & 0 & 2 \\ 0 & 1 & -1 \\ 0 & 0 & 1 \end{bmatrix}$$

$$\mathbf{R} = \begin{bmatrix} -2/\sqrt{8} & -1/\sqrt{2} & 0 \\ -2/\sqrt{8} & 1/\sqrt{2} & 0 \\ 0 & 0 & 1 \end{bmatrix}$$

$$\mathbf{S} = \begin{bmatrix} \sqrt{8} & 0 & 0 \\ 0 & \sqrt{2} & 0 \\ 0 & 0 & 1 \end{bmatrix}$$

To check our answer, we can work out the expression **TRS** and it should result in matrix **M**. Note that the order is **TRS** because we extracted the column scalars of **M** (which is the standard approach). If we had extracted the row scalars, our order would be **SRT**.

M = TRS

$$\mathbf{M} = \begin{bmatrix} 1 & 0 & 2 \\ 0 & 1 & -1 \\ 0 & 0 & 1 \end{bmatrix} \begin{bmatrix} -2/\sqrt{8} & -1/\sqrt{2} & 0 \\ -2/\sqrt{8} & 1/\sqrt{2} & 0 \\ 0 & 0 & 1 \end{bmatrix} \begin{bmatrix} \sqrt{8} & 0 & 0 \\ 0 & \sqrt{2} & 0 \\ 0 & 0 & 1 \end{bmatrix}$$

$$\mathbf{M} = \begin{bmatrix} 1 & 0 & 2 \\ 0 & 1 & -1 \\ 0 & 0 & 1 \end{bmatrix} \begin{bmatrix} -2 & -1 & 0 \\ -2 & 1 & 0 \\ 0 & 0 & 1 \end{bmatrix} = \begin{bmatrix} -2 & -1 & 2 \\ -2 & 1 & -1 \\ 0 & 0 & 1 \end{bmatrix}$$

You can view a worked solution of this problem on Stack Exchange [5].

MODEL VIEW PROJECTION MATRICES

In the earlier inverse matrix section, we spent a lot of time establishing why inverses are "no bueno" (that's Spanish for "not good"). Why would we even want to compute them? Model view projection (MVP) matrices are a great example to motivate our study. MVP refers to three matrices which are used to transform every point to the screen and it's a key step of graphics programming. The basic idea here is that every point is multiplied through each of these matrices and with each computation the point is transformed to a new space.

Here's an overview on what each of the three matrices represent:

MVP = (Local **M**odel to World) (World to **V**iew) (View to **P**rojection Screen)
Model matrix (per model): defines position, rotation, and scale of model vertices in-world.
View matrix (per camera): defines position and orientation of the camera.
Projection matrix (per camera): Maps what camera sees to normalized device coordinates (NDC) where (-1,-1) typically represents the bottom-left of the screen. It also handles aspect ratio and perspective.

Even if you haven't done explicit graphics programming, you're probably familiar with the conceptual purpose of the model matrix, as many transform APIs present the ability to set a transform's global or local position. For example, in Unity you can set an object's position using `transform.position = value` (which implicitly sets the 'global' position) or by using `transform.localPosition = value`. When we set a transform's local position, we are simply altering its transform matrix directly. If we refer back to our RBTM diagram in Figure 8.2, we are setting the blue 1×3 translation vector. It may seem odd that our position is referred to as a "translation" but that is because the position (1, 1, 1) can be thought of as a translation of [1, 1, 1] from the origin. And while it wouldn't make sense to apply multiple positions to an object, it does make sense to apply multiple translations, which is what happens in a transform hierarchy. In a transform hierarchy, an object can be stored as a child object of another object. The effect of this setup is that any points for the child object are multiplied through the child's transform matrix and then its parent matrix, grandparent matrix, and so on.

> **Practice Problem 8.7**: EntityA's global position is (50, 50), and EntityB's is (80, 90). What is the position of EntityB in EntityA's local space?
>
> > **Hint**: Another way to think of this might be… if we made EntityB a child of EntityA, what would the local position of EntityB be?
> > **Solution**: If EntityA is at (50, 50) then we need to travel [30, 40] to get to (80, 90) so EntityB's local position would be (30, 40) in EntityA's space.

WORKS CITED

[1] Wikimedia Commons Contributors, "Matrix multiplication diagram 2," *Wikimedia Commons*, Oct. 4, 2010. [Online]. Available: https://en.m.wikipedia.org/wiki/File:Matrix_multiplication_diagram_2.svg. [Accessed: Aug. 17, 2024].

[2] S. Khan, "Linear Transformation Examples: Rotations in R2," *Khan Academy*. [Online]. Available: https://www.khanacademy.org/math/linear-algebra/matrix-transformations/lin-trans-examples/v/linear-transformation-examples-rotations-in-r2. [Accessed: Aug. 17, 2024].

[3] Leios Labs, "What are affine transformations?" *YouTube*. [Online]. Available: https://www.youtube.com/watch?v=E3Phj6J287o&ab_channel=LeiosLabs. [Accessed: Aug. 17, 2024].

[4] Rodriguez, J. (2023). *Math for Game Developers - Faster Matrix Inversions* [Video]. YouTube. https://youtu.be/7CxKAtWqHC8

[5] metavers, "Given this transformation matrix, how do I decompose it into translation, rotation and scale matrices?," *Mathematics Stack Exchange*, Aug. 23, 2013. [Online]. Available: https://math.stackexchange.com/questions/237369/given-this-transformation-matrix-how-do-i-decompose-it-into-translation-rotati. [Accessed: Aug. 18, 2024].

Maths
Rotations

9

ANGLES

Roll, Pitch, Yaw

Rotations are usually expressed as three angles representing rotations along the roll, pitch, and yaw axes. You should be familiar with the rotation type that each of these terms refers to. I've included an example with each term, but please remember that they depend on which direction the axis points. In this case, I am assuming x is "forward", y is "right" and z is "up", which may not always be the case.

- Roll usually means an x-axis rotation. Imagine a dog rolling over to its left or right.
- Pitch usually means a y-axis rotation. Imagine an airplane pitching up or down.
- Yaw usually means a z-axis rotation. Imagine a person twisting their head sideways as they yawn.

Usually roll, pitch, and yaw correspond to x, y, z respectively. To remember the order "roll, pitch, yaw" you can use the mnemonic phrase "real programmers YouTube".

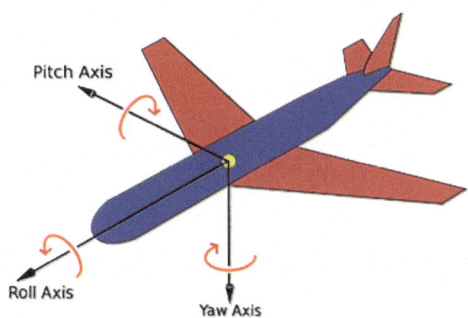

FIGURE 9.1 Demonstration of roll, pitch, and yaw [1].

DOI: 10.1201/9781003565550-11

Triangle Angle Derivations

Triangles are at the core of most vector maths problems we will solve. In particular, right triangles (where one of the three angles is 90°) have a ton of incredibly convenient and reliable properties that can help us. Be very familiar with the SoCahToa formulae (Figure 9.2) which, in the case of right triangles, allows us to quickly derive an angle as a ratio of two triangle side lengths.

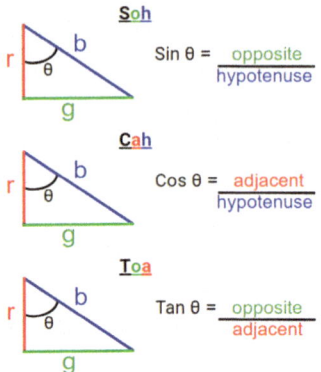

FIGURE 9.2   SohCahToa for triangle angles.

Often triangles will not be right triangles, or we will only have two sides of a triangle defined (via two vectors). In these cases, it helps to have some backup approaches for deriving our internal angles. Given two vectors, recall that the dot product is one way to derive the angle.

$$\theta = \operatorname{acos}((\mathbf{a} \cdot \mathbf{b}) / (\| \mathbf{a} \| * \| \mathbf{b} \|) * (180 / \pi)$$

In rare cases when you don't know the two vector lengths, there is an atan2 method that we can use which requires the cross product.

$$\theta = \operatorname{atan2}\left(\operatorname{abs}(\| \mathbf{a} \times \mathbf{b} \|), \mathbf{a} \cdot \mathbf{b} \right) * (180 / \pi)$$

The above conversion uses the atan2 function which may be unfamiliar to you. atan2(y, x) does the exact same thing as atan(y/x) but does so in such a way that the output tends to be from -180° to 180° rather than 0° to 360°. This is a really cool function because it can tell us the theta of a rotation represented by the point xy as shown below. We'll see atan2 come up again when we explore quaternions.

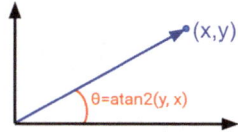

FIGURE 9.3 atan2 can calculate the angle it takes to reach a point.

When working with angles it also useful to know that you can transition between sine and cosine like this:

$$\sin(\theta) = \cos(90° - \theta)$$

$$\cos(\theta) = \sin(90° - \theta)$$

Triangle Side Length Derivations

Here's another perspective on SohCahToa, focused on deriving side lengths. In this case, if you have a right triangle and want to determine the length of a non-hypotenuse side, you can multiply the hypotenuse length with cosine of the adjacent angle or sine of the opposite angle. Deriving the side lengths of a triangle in this way can help you determine the angles of a rotation. You can usually think of a rotation as a triangle, with two of its sides formed by vectors aimed in the direction of the start and end orientation.

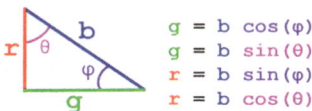

FIGURE 9.4 SohCahToa for triangle side lengths.

Pythagorean Theorem

You should also memorize the Pythagorean theorem which states that the squared length of a right triangle's hypotenuse is equal to the sum of each other side's squared length:

$$a^2 + b^2 = h^2$$

45-45-90 Theorem

In the specific case of an isosceles right triangle, the side lengths will always form the ratio 1:1: √2 and the angles will be 45°, 45°, 90°.

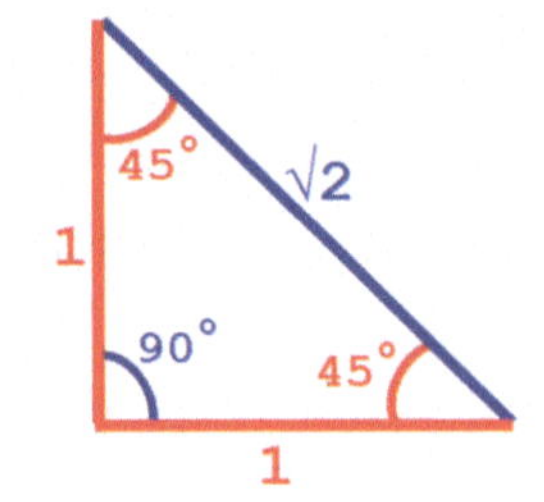

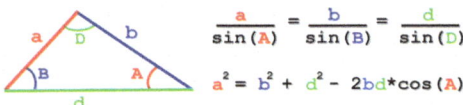

FIGURE 9.5   45-45-90 Theorem.

Sine Rule and Cosine Rule

Lastly, I would recommend learning the Sine Rule and maybe the Cosine Rule as well. They are just additional ways to relate triangle side lengths with angles, but what's nice about them is they don't require right triangles to work.

$$\frac{a}{\sin(A)} = \frac{b}{\sin(B)} = \frac{d}{\sin(D)}$$

$$a^2 = b^2 + d^2 - 2bd*\cos(A)$$

FIGURE 9.6   Sine Rule and Cosine Rule.

BARYCENTRIC COORDINATES

The barycentric coordinate system is used primarily to describe points within a triangle. The idea behind them is that if we split a mass such that each of the triangle's vertices has a different portion of the mass, where would the overall triangle's center of mass be?

For example, if we put all of the mass on the triangle's first vertex and zero mass on the others, then that might be expressed as (1.0, 0, 0). This would result in a center of mass on the first vertex and would represent a coordinate for that position. Likewise, we could express the triangle's second vertex as (0, 1.0, 0) or its third as (0, 0, 1.0). A point halfway between the edges of the first two points would have coordinates (0.5, 0.5, 0). I should also note that barycentric weights can be negative, in which case the resulting point lies outside of the triangle.

Barycentric Coordinates - Geometric Derivation

If a point lies in the triangle, its Barycentric coordinates are equivalent to the ratio of areas formed by the inner triangles formed by that point. As shown in Figure 9.7, the coordinate associated with each vertex is calculated by dividing the area of the inner triangle opposite that vertex with the sum of all the inner triangle areas.

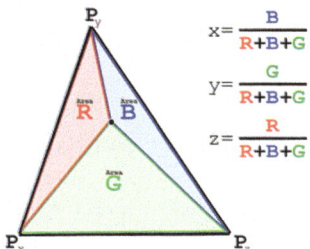

$$x = \frac{B}{R+B+G}$$

$$y = \frac{G}{R+B+G}$$

$$z = \frac{R}{R+B+G}$$

FIGURE 9.7 Barycentric coordinates, geometric derivation.

Barycentric Coordinates - Algebraic Derivation

While calculating Barycentric coordinates via the geometric approach is intuitive, it's not performant and it's unable to express points that lie outside of the triangle. The algebraic derivation provides us with the same coordinates through an expression that incorporates the triangle's vertices with several dot products.

1. First, we define vectors between the triangle vertices:
   - $v_0 = B - A$
   - $v_1 = C - A$
   - $v_2 = P - A$
2. Then we precompute several dot products between these vectors:
   - $d_{00} = v_0 \cdot v_0$
   - $d_{01} = v_0 \cdot v_1$
   - $d_{11} = v_1 \cdot v_1$
   - $d_{20} = v_2 \cdot v_0$
   - $d_{21} = v_2 \cdot v_1$
3. We precompute the shared denominator:
   - Denominator $= d_{00} * d_{11} - d_{01}^2$
4. And finally, we express the three coordinates in terms of the dot products and denominator:
   - $x = (d_{11} \cdot d_{20} - d_{01} \cdot d_{21})$ / Denominator
   - $y = (d_{00} \cdot d_{21} - d_{01} \cdot d_{20})$ / Denominator
   - $z = 1 - x - y$

Practice Problem 9.1: Determine if point P lies within the triangle defined by points ABC.

Solution: This is a very common problem, maybe one of the most common problems in all computer graphics. So, there are very many ways to solve it.

If you're comfortable with Barycentric coordinates, then determining those for point P is a viable approach. As long as none of the coordinates are negative, P lies within ABC.

A more efficient approach is to use the 2D cross product discussed in a previous chapter on vectors. The 2D cross product can determine what side of a line a point is on. If you test P against each edge of ABC and it is on the same side of all edges, then P lies within ABC. Not only is this approach very efficient, but it actually works for all convex polygons and similar approach considering faces instead of edges works in the 3D case for all convex polyhedrons.

Radians

Radians are a unit to measure angles and are interchangeable with degrees. You should feel comfortable converting increments of 45° with their radian counterparts.

Degrees	0°	45°	90°	135°	180°	225°	270°	315°	360°
Radians	0π	π/4	π/2	3π/4	π	5π/4	6π/4	7π/4	2π

Additionally, here are two simple conversion formulae, just plug in $\underline{x}$ for result $\underline{y}$.

$$\underline{x} \text{ radians} * (180 / \pi) = \underline{y} \text{ degrees}$$
$$\underline{x}° * (\pi / 180) = \underline{y} \text{ radians}$$

UNIT CIRCLE

When we take the cosine and sine of an angle, we are calculating the (x, y) coordinate of that angle on the unit circle. The diagram below is called the unit circle because its radius is one unit. It's important that you feel comfortable with this diagram. It is used to convert angles to their corresponding x and y coordinates.

As an example, if we have an angle of π/2 radians (90°) we can imagine it on the circle. Our angle starts from 0°, then grows along the pink line of Figure 9.8 until it spans the whole top-right quadrant. If the unit circle was a clock, 90° would represent the clock hand going from "3", counterclockwise to "12". In this position, at the top of the circle, the coordinate is written as "(0, 1)" which represents that cos(90°) = 0 and sin(90°) = 1.

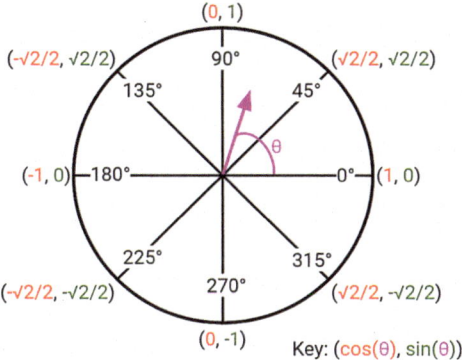

FIGURE 9.8   The Unit Circle.

Practice Problem 9.2: Simplify the expression: sin(π).

Solution: π is 180°, and I can see that the unit circle coord of 180° is (-1, 0). Sin takes the y value which is 0. Therefore sin(π) = 0.

POLAR COORDINATES

Similar to how radians are a way to express angles as an alternative to degrees, polar coords are a way to express positions as an alternative to Cartesian coordinates (the standard "x,y,z"). In the below example image, the red position can be expressed in cartesian form as (x, y) or polar form as (r, θ). Polar coordinates are particularly useful for describing circular movement relative to a center point, such as around a sphere.

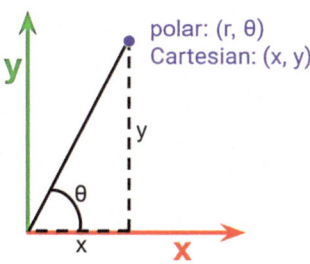

FIGURE 9.9   Polar coordinates.

QUATERNIONS

We have previously discussed rotation matrices which are often used with Euler angles to represent rotational data. But Euler angles cannot be used for smooth interpolation because of gimbal lock. As an alternative to Euler angles, we have quaternions which use imaginary numbers to represent rotations through a 4D hypersphere instead of in a 3D context to prevent gimbal lock. If you are not familiar with imaginary numbers, I recommend the YouTube video "Introduction to i and imaginary numbers | Imaginary and complex numbers | Precalculus | Khan Academy" by Khan Academy [2].

By far, one of the best courses I have ever taken is Ben Eater's and 3Blue1Brown's "Visualizing quaternions" course [3]. It's a free course that has a ton of interactive demos which are always the sort of exercises I encourage my students to partake in. There's currently no better way that I know of to gain an intuition on what quaternions mean. Please, if you have not taken their course already, set aside an evening to do so.

A quaternion's formula:

$$\mathbf{q} = \left(w, \hat{v}\right) = \left[\cos\left(\theta/2\right),\ \hat{v}_x\ \sin\left(\theta/2\right),\ \hat{v}_y\ \sin\left(\theta/2\right),\ \hat{v}_z\ \sin\left(\theta/2\right)\right]$$

- θ is the angle of rotation.
- $\hat{v}$ is a unit vector representing the axis of rotation.
- $w = \cos(\theta/2)$ is the scalar part.
- $\hat{v} \sin(\theta/2)$ is the vector part.

A quaternion is a type of complex number with four parts: one real number (w) and three imaginary numbers (x, y, z). You can think of it as an extension of complex numbers to higher dimensions. A quaternion's imaginary parts (x, y, z) form an axis to define the direction of rotation and w (which ranges from 0 to 1) defines the amount of rotation.

- Rotation quaternions are "unit" quaternions which means that the sum of the squares of their four parts equals 1. The reason why it's the "sum of squares" rather than just the sum like normal unit vectors is because in order to apply a quaternion you actually multiply with it twice. So, it's the sum of squares that represents the total distortion of whatever it is affecting.
- Quaternions can represent all rotations in 3D space in two ways. For every rotation, there are two quaternions that describe it: one for the rotation and one for the opposite direction. This is called a "double cover," and it's part of the reason why quaternions do not suffer from gimbal lock.

Practice Problem 9.3: Construct a quaternion to rotate 270°, counterclockwise, around the z-axis.

Solution: First, we define our axis of rotation. In a quaternion that is completely handled by the imaginary xyz part. In this case, it's simple, since we are rotating around the literal z-axis our z value will be 1, with x and y as 0. Next, we just need to plug in the amount of rotation. In a quaternion construction, the total rotation is split between sine and cosine evenly. So in this case, since our total rotation is 270°, that means our real part w will be $\cos(270° / 2) + \sin(270° / 2)$.

The result is $\left(\cos\left(135°\right), \sin\left(135°\right)\left[0i + 0j + 1k\right]\right)$.

You can view a worked solution of this problem in the aforementioned "Visualizing quaternions" course in the "Double Cover" lesson [4].

Quaternion Maths

To apply a quaternion **q** to point P, you multiply it on both sides. On the left you place the quaternion itself, and on the right, you place the quaternion's inverse: $q \times P \times q^{-1}$. The **q** term deforms the hyper-sphere, but q^{-1} puts it back in place

Slerp vs Lerp vs NLerp

Linearly interpolating quaternions is not very straightforward. That was a joke, but it's also true because following a linear path, rather than the surface of the 4D hyper-sphere, will result in a faster rotation, generally appearing less "smooth". While this probably is not noticeable for small rotations, generally Slerp or NLerp is used rather than Lerp. There's an Inner Product article on this topic titled "Understanding Slerp, Then Not Using It" by Jonathan Blow that I recommend checking out for additional information [5].

Quaternion from Euler

Quaternions, compared to Euler angles, are just another way to deal with the same underlying data. We can freely switch between the two systems. Below is what the code looks like to construct a quaternion from Euler angles. Notice in the code how we take the half angles, such as "pitch * 0.5f". Remember, we do this because a quaternion is always multiplied twice when it's applied.

```
Quaternion ToQuaternion(float yaw, float pitch, float roll) {
    float cosYaw = cos(yaw * 0.5f);
    float sinYaw = sin(yaw * 0.5f);
    float cosPitch = cos(pitch * 0.5f);
    float sinPitch = sin(pitch * 0.5f);
    float cosRoll = cos(roll * 0.5f);
    float sinRoll = sin(roll * 0.5f);
    Quaternion q;
    q.w = cosRoll * cosPitch * cosYaw + sinRoll * sinPitch * sinYaw;
    q.x = sinRoll * cosPitch * cosYaw - cosRoll * sinPitch * sinYaw;
    q.y = cosRoll * sinPitch * cosYaw + sinRoll * cosPitch * sinYaw;
    q.z = cosRoll * cosPitch * sinYaw - sinRoll * sinPitch * cosYaw;
    return q;
}
```

Likewise, we can convert any quaternion's four values back into its Euler angles as shown below:

$$\begin{bmatrix} a, b, c, d \end{bmatrix} \rightarrow \begin{bmatrix} x, y, z \end{bmatrix}$$

$$x = \operatorname{atan2}\left(2(ab + cd), 1 - 2(b^2 + c^2) \right)$$

$$y = \operatorname{asin}\left(2(ac - db) \right)$$

$$z = \operatorname{atan2}\left(2(ad + bc), 1 - 2(c^2 + d^2) \right)$$

WORKS CITED

[1] Auawise, "Yaw Axis.svg," Wikimedia Commons, Jul. 31, 2008. [Online]. Available: https://commons.wikimedia.org/wiki/File:Yaw_Axis.svg. [Accessed: Aug. 18, 2024].

[2] Khan Academy, "Introduction to i and imaginary numbers I Imaginary and complex numbers I Precalculus I Khan Academy," *YouTube*, Aug. 15, 2016. [Online]. Available: https://www.youtube.com/watch?v=ysVcAYo7UPI&ab_channel=KhanAcademy. [Accessed: Aug. 17, 2024].

[3] G. Sanderson and B. Eater, *Visualizing quaternions: An Explorable Video Series*. [Online]. Available: https://eater.net/quaternions. [Accessed: Aug. 17, 2024].

[4] G. Sanderson and B. Eater, *Visualizing Quaternions: An Explorable Video Series*. [Online]. Available: https://eater.net/quaternions/video/doublecover. [Accessed: Aug. 17, 2024].

[5] J. Blow, "Understanding Slerp, Then Not Using It," Number None, Available: http://number-none.com/product/Understanding%20Slerp,%20Then%20Not%20Using%20It. [Accessed: Aug. 18, 2024].

Maths
Planes

10

Planes are 2D structures that stretch out into infinity and are therefore the most "generalized" 2D structure. If you can learn how to solve a generalized problem, then you'll unlock an understanding of how to solve a whole class of problems like it. For example, earlier in the vectors chapter, we discussed how a line is a generalized 1D structure. Rays and line segments are just different restrictions on the line's generalized form. When you go about solving if two rays intersect, you approach the problem by first solving the generalized problem of determining if the lines which contain those rays intersect. Once you've found the point of intersection, if any, you can then solve the specific problem by testing if that point is within the bounds of your specific rays. This is an important paradigm to recognize. Triangles are planes with restrictions. Just like with the rays, if you get a triangle intersection problem, you first approach the generalized problem of testing if the plane containing the triangle has an intersection. And if it does, you can then test if the intersection is within the triangle. Try to keep this in mind as you learn about planes, because planes are the "generalized" form of all polygons; so you're really learning about a very broad class of problems.

Below is the formula for a plane:

$$ax + by + cz = d$$

- **n** is a vector formed by [a, b, c], representing the norm of the plane.
- d / |**n**| is the signed distance from the plane to the origin.

In the above equation, d is the positive distance from the origin. Sometimes you will see the equation in this form $ax + by + cz - d = 0$. In those cases, d is the negative distance from the origin.

The norm of the plane, usually represented by vector **n**, is the plane's "normal vector" which means that it is perpendicular to any line within the plane. Usually, the plane's normal vector is normalized in which case we will write it as *n*. But unless otherwise stated during an interview, you should assume that a plane's normal is not normalized.

Planes can be represented in many ways, including:

- $ax + by + cz = d$, the positive distance formula.
- $ax + by + cz - d = 0$, the classical, negative distance formula.
- Three points within the plane (a triangle), as long as they are not all on the same line.
- One point and two vectors within the plane (parametric form).
- One point "P" on the plane, and the plane's normal "**n**". In this case, we can express a way to test if an input point "X" is on the plane using the following equation: $\mathbf{n}\cdot(X-P)=0$, which translates to "any line segment constructed between two points on the plane (X and P) will be perpendicular to the plane's normal (**n**)." After distributing the dot product, you can rewrite the equation as $\mathbf{n}\cdot X = \mathbf{n}\cdot P$. And if you expand $\mathbf{n}\cdot P$, you get the expression: $ax + by + cz$ which is half of the classical plane formula.

Practice Problem 10.1: Consider the formula for a plane: $ax + by + cz = d$. Consider if any one of these values (a, b, c, or d) were 0, what would that mean geometrically for the plane?

Solution: This is a great question to test your familiarity with the formula's components.

- If $d == 0$, then the plane passes through origin.
- If $a == 0$, then the plane extends parallel to the x-axis, meaning there is no constraint on that axis.
- If $a == 0$ and $b == 0$, then the plane is comprised of lines parallel to the x-axis and y-axis, and the plane's norm is parallel to the z-axis.

Given two planes expressed in the form $ax + by + cz = d$, we can compare the ratios of each plane's terms to understand how the planes intersect. For example, the top row means that if the ratios of both planes' a, b, and c terms are not all equivalent, then the planes must be intersecting.

If $\dfrac{a_1}{a_2} \neq \dfrac{b_1}{b_2} \neq \dfrac{c_1}{c_1}$, then planes are intersecting.

If $\dfrac{a_1}{a_2} = \dfrac{b_1}{b_2} = \dfrac{c_1}{c_1} \neq \dfrac{d_1}{d_2}$, then planes are parallel.

If $\dfrac{a_1}{a_2} = \dfrac{b_1}{b_2} = \dfrac{c_1}{c_1} = \dfrac{d_1}{d_2}$, then planes are coincident.

PRACTICE PROBLEMS

Practice Problem 10.2: Given a plane defined by a, b, c, and d, derive a point "P" that lies on the plane.

Hint: The plane's d value describes its distance from the origin. And its norm describes the shortest path from any structure (including perhaps a point, like the origin) to the plane.

Solution: First let's construct the norm of the plane using a, b, and c. $\mathbf{n}$ = [a, b, c]. Then normalize $\mathbf{n}$ by dividing it by its own magnitude. $\hat{n}$ = $\mathbf{n}$ / ||n||. This $\hat{n}$ is the normalized normal, a very useful vector. Since the shortest distance between two structures is a straight line, that means the shortest distance from any point to this plane is along $\hat{n}$.

Remember, d / ||n|| is the signed distance from origin. So, if we calculate this value: d' = d / ||$\hat{n}$||, Then d' is our distance from the origin to the plane.

Now if we scale $\hat{n}$ by d', and add it to the origin, we will get a point on the plane. And not just any point, but the point on the plane <u>closest</u> to the origin. I like to think of this point as the 'center' of the plane but that is kind of silly since a plane is infinite and therefore doesn't really have a center. Here's the final equation: P = (0,0,0) + ($\hat{n}$ * d').

Note: A common mistake on this problem is to choose arbitrary x and y values (such as x=0 and y=0) and then plug them in to solve for z. This approach is invalid because it assumes the plane contains these values for x and y. This is not a valid assumption, for example the plane may be perpendicular to the x axis but intersect it at x=5.

Practice Problem 10.3: Given a plane defined by a, b, c, and d, and a separate point, P_1, determine if P_1 is above, below, or on the plane.

Solution: It's easy to check if P_1 is on the plane, just plug it into the formula.

$$ax + by + cz = d$$

$$a\left(P_1.x\right) + b\left(P_1.y\right) + c\left(P_1.z\right) = d$$

If both sides of the equation are not equal, then P_1 is not on the plane.

If P_1 is not on the plane, then we are going to need a point on the plane, P_{plane}, to proceed. Solve the earlier practice problem to find a point on the plane if you do not have one already.

Construct vector $p = P_{plane} - P_1$, this is a vector that extends from our starting point to the plane. Now we can do a dot product between this vector and the plane's normal. From the dot product, we can determine the angle between these vectors. If the vector is acute, then we know the point is on the side of the plane that the normal points toward, which is generally regarded as the "top" of the plane or "above" the plane. If the angle is obtuse then the point is below the plane. The angle will not be 90° because we already determined that the point is not on the plane.

Practice Problem 10.4: Given a plane defined by a, b, c, and d, and a separate vector **v**, project **v** onto the plane.

Hint: Any vector can be broken into component vectors. For example the vector [x, y, z] can be broken into [x, 0, 0] + [0, y, 0] + [0, 0, z]. In this problem we want the component of **v** that is parallel to the plane. This is the same as saying that we want **v** other than its components not parallel to the plane.

Solution: In 3D, a plane is a hyperspace. What's a hyperspace? Well, if we are dealing with an n-dimensional subspace then a hyperspace is any n-1 dimensional subspace. In this case we are in 3D, aka R^3, so n = 3. A plane is a 2-dimensional subspace, so a plane is a hyperspace of R^3.

This matters because if we remove any components of **v** that <u>are not</u> in the plane, we will be left with the components of **v** that <u>are</u> in the plane. To first get the part of **v** that is not in the plane, project **v** onto the plane's normal. Let's call this **v**'. Now we just remove **v**' from **v** which is as easy as **v** − **v**' = **v**″ and that's our solution.

Practice Problem 10.5: Given a plane defined by a, b, c, and d, and a separate point, P_1, find the closest point on the plane to P_1.

Hint: This is the kind of question where if you are stuck you should definitely try drawing a picture. Give that a try and see if you can "connect the dots" from there.

Hint: So, you've now drawn a picture. Let's say it includes the plane and point P_1 somewhere above the plane. Our goal is to define some P_2 on the plane, draw it. The shortest distance between two structures is going to be a straight line. So, draw a vector from P_1 to P_2 along the plane's normal; since this is something we need to derive, I recommend drawing it as a dotted line. While we cannot derive P_2 immediately, we can derive a point on the plane to get us to P_2. Let's call this intermediate point $P_{1.5}$. How can we form a triangle between our two known points, P_1 and $P_{1.5}$, in order to derive our unknown goal point, P_2?   .

Solution: Find any point on the plane, call it $P_{1.5}$. Project line segment $P_1P_{1.5}$ onto the plane and it will point from $P_{1.5}$ to P_2.

Practice Problem 10.6: Given a plane defined by a, b, c, and d, and a separate point P_1, find a line that is parallel to the plane and contains P_1.

Hint: Draw a picture, find the right triangle, and solve derive the missing side.

Solution: Find a point on the plane, call it P_2. Project $P_1 P_2$ onto the normalized norm of the plane and it will point to P_3. P_1P_3 is parallel to the plane and contains P_1.

Practice Problem 10.7: Given a plane defined by a, b, c, and d, and a separate point P_1, find the distance from the point to the plane.

Solution: Find the normalized norm of the plane, call it $\hat{n}$. Find any point on the plane, call it P_2. Project P_2P_1 onto $\hat{n}$. The resulting vector, $\mathbf{v}$, is the shortest vector from the plane to the point. Calculate the magnitude of the vector for the distance.

Practice Problem 10.8: Given a plane defined by a, b, c, and d, and a separate line JK, determine their intersection.

Solution: The first insight is that if JK is not parallel to a plane, then it must be intersecting the plane. This is because a plane is defined by two orthogonal vectors and if we add a third vector then we will have a basis in R^3 (3-dimensional space). Remember that your set of vectors does not need to be orthogonal to be a basis, they only need to be non-collinear. For example, [1,0] and [1,1] form a basis for R^2.

So our first test is to determine if JK is parallel to the plane, can you remember how to determine that? Here's a hint: in order to see if a line is parallel to the plane, that's the same as checking if the line is perpendicular to something perpendicular to the plane. We can take the plane's normal vector, and if it's perpendicular to JK, then JK is parallel to the plane. Calculate the dot product of JK and the plane's normalized norm...

If the dot product result is zero, then the line is parallel to the plane. So that means we can have a coincident case (no intersections) or a disjoint case (infinite intersections, because JK is on the plane). To determine which of these cases we have, take a point from the line and see if it satisfies the plane's formula.

If the result of the dot product was non-zero, then we know JK intersects the plane, and we just need to find the intersection point. Convert the plane and line into their parametric equations and combine them to solve for a point.

$$(X - Q) \cdot \hat{n} = 0$$

This is the plane parametric formula where Q is a point on the plane and X is an input point.

$$P + t\mathbf{d}$$

This is the parametric formula for a line, where P is a start point and $\mathbf{d}$ is the direction vector.

Let us now combine these formulas to solve for t, the "time" of intersection.

$$(X - Q) \cdot \hat{n} = 0$$
$$((P + t\mathbf{d}) - Q) \cdot \hat{n} = 0$$
$$(P - Q + t\mathbf{d}) \cdot \hat{n} = 0$$
$$(P - Q) \cdot n + t\mathbf{d} \cdot \hat{n} = 0$$

Now we subtract "(P - Q) · n" from both sides…

$$t\,\mathbf{d}\cdot\hat{\boldsymbol{n}} = (Q-P)\cdot\hat{\boldsymbol{n}}$$
$$t = (Q-P)\cdot\hat{\boldsymbol{n}}\,/\,(\mathbf{d}\cdot\hat{\boldsymbol{n}})$$

Aha, we have solved for t! But beware, there is a division here so we should be careful not to divide by zero. The denominator will only be zero when the plane normal and the JK have a dot product of zero. And as you will recall, we earlier made sure that was not the case, so we can rest assured that we will not have a zero denominator.

If this line was instead a ray or line segment, we would just need to verify that our t result lies within a valid range for the structure. To determine the exact intersection point, we can plug this t back into the parametric formula for a line.

Practice Problem 10.9: Given a plane defined by a_1, b_1, c_1, and d_1, and another plane defined by a_2, b_2, c_2, and d_2, determine their intersection.

Hint: First consider how to test if the planes are parallel and what edge cases that situation would cover. If the planes are not parallel, their intersection will be a line. Here's a diagram to illustrate the line of intersection.

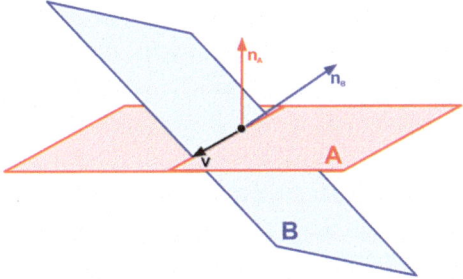

FIGURE 10.1 **v** defines the line of intersection.

Solution: Similar to the last problem, we first want to determine if these planes are parallel. Derive the plane norms n_1 and n_2. If the planes are parallel, then their normals will also be parallel. To check if the normals are parallel, we can use the cross product and check if the result is the zero vector: $\hat{\boldsymbol{n}}_1 \times \hat{\boldsymbol{n}}_2 = \mathbf{v}$.

If the resulting vector **v** is **0**, test a point to see if the planes are coincident. If they are, then the planes are coincident and either of them can be used to define their intersection. If they are not coincident, then they are parallel, aka disjoint, and have no intersection.

If the resulting vector **v** is not **0**, then **v** defines the direction vector of the line of intersection between the two points.

But we are not done, since a direction vector is only half of the information we need to define a line. The other half is at least one point. So now we just need to determine a point that lies on both planes. Sadly, between all of each plane's variables, we have too many unknowns to solve this problem outright. But since we now have **v**, we can use it to simplify the problem. Any component of **v** that is non-zero can be used to safely zero the corresponding components of the plane formulas.

For example, if our plane formulas are…

$$a_1x + b_1y + c_1z - d_1 = 0 \text{ and } a_2x + b_2y + c_2z - d_2 = 0$$

Then if the **v**.z != 0, we can zero the z components of our formula:

$$a_1x + b_1y + c_1 * 0 - d_1 = 0 \text{ and } a_2x + b_2y + c_2 * 0 - d_2 = 0$$
$$a_1x + b_1y - d_1 = 0 \text{ and } a_2x + b_2y - d_2 = 0$$

If **v**.z was zero then we would have chosen the x or y component instead. We know that at least one component is not zero because the earlier part of this problem was determining that **v** != 0.

Once we have simplified the formulas, we can combine them to express x and y using only the terms of the planes.

$$x = \left(b_1 * d_2 - b_2 * d_1\right) / \left(a_1 * b_2 - c_1 * b_1\right)$$
$$y = \left(b_2 * d_1 - a_1 * d_2\right) / \left(a_1 * b_2 - c_1 * b_1\right)$$
$$z = 0$$

We solve to determine point P = [x, y, z].

Then we have completed the problem and determined the intersection to be the line defined by **v** and P.

Maths
Polygons and Polyhedrons

11

POLYGONS

Polygons are 2D shapes. A triangle is a polygon with three sides, a rectangle is a polygon with four sides, and so on. Remember that it is often useful to think of polygons as restricted planes. Many problems, such as "does the line intersect the polygon", are easy to solve if we simplify them to "does the line intersect the plane which contains the polygon". Once we solve that generalized problem, we can always add the specificity back in to ask, "given the intersection on the polygon's plane, does it lie within the polygon's actual bounds?"

Convexity vs Concavity

A 2D or 3D object is convex if you can connect any two of its points in order to form a line segment that is entirely contained within the object. Basically, convexity means no holes or caves. As a mnemonic, you can think of con<u>cave</u> polygons as those with "caves". The classic platonic solids are all convex. A convex polygon's internal angles will all be less than 180°, a concave polygon will have at least one internal angle that is more than 180°.

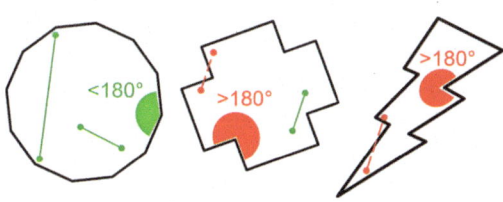

FIGURE 11.1   A concave polygon contains at least one internal angle that is more than 180°.

DOI: 10.1201/9781003565550-13

POINT VS POLYGON/POLYHEDRON INTERSECT

A very common interview problem will ask candidates to determine if a point is contained within a convex polygon or polyhedron. For these problems, I recommend an approach that considers each edge or face as a constraint and checks one-by-one if the point meets all of the constraints. There are considerable optimizations that can be made if we have additional information about the object, but this approach is simple to remember and can be used for both 2D and 3D.

Practice Problem 11.1: Is point M in the rectangle defined by points A, B, C, and D?

Solution: There are many ways to solve this problem, here are a few approaches with increasing efficiency.

Approach #1: The first approach I will outline is geometric. It's simple to conceptualize but its performance is slow. Imagine M indeed lies within rectangle ABCD. If that is the case, then when can use the rectangle's vertices with M to form four triangles: ABM, MBC, DMC, and AMD.

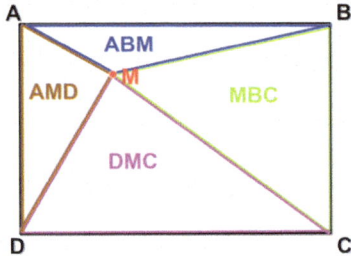

FIGURE 11.2   Four triangles formed by M and the vertices of rectangle ABCD.

Sum the area of each triangle and compare it to the rectangle ABCD's total area. If the sum of triangle areas is greater than the rectangle's total area, then M must be outside of the rectangle. If the sum is equal to or less than the total area, then M is within ABCD. If the sum is equal to the total area, then we check if one triangle had zero area which would indicate that M lies on an edge of ABCD. You can generalize this approach to any convex polygon for which you can easily calculate its area. The number of triangles you'll need is equal to the number of sides.

Approach #2: The second approach we will cover is trigonometric. In this approach we will use the 2D cross product to determine if the point is on the "left" or "right" side of each of the rectangle's edges. As long as we are consistent in checking the edges with a clockwise direction, it is sufficient to check if the point is on the same side of every edge. Here's a diagram to visualize the process.

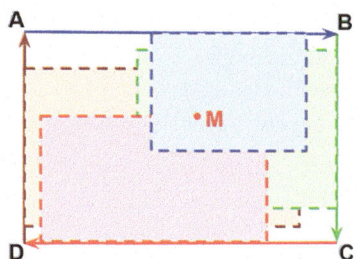

FIGURE 11.3   We check if the point is the area to the right of each line.

In Figure 11.3 I drew each edge going clockwise and you can see that, if they all go clockwise, the right-hand side of each edge extends into the center of the rectangle. By checking that M is on the right-hand side of each edge, we are checking if it is within the rectangle. This also illustrates why it is important for our edge construction to be consistent, because if the edges all went counterclockwise, we would need to check if the point was on each of their left-hand sides instead. And if some edges were clockwise and some were counterclockwise, then we could not apply the same check to all edges. The 2D cross product is applied as follows...

In order to test whether the point (X_P, Y_P) lies to the left or right of the edge $(X_A, Y_A) - (X_B, Y_B)$, you just need to calculate $D = (X_B - X_A) * (Y_P - Y_A) - (X_P - X_A) * (Y_B - Y_A)$. If $D > 0$, the point is on the left-hand side. If $D < 0$, the point is on the right-hand side. If $D = 0$, the point is on BA.

This approach works for all convex polygons, whether they have three or three hundred sides!

Approach #3: While approaches like the previous one are helpful to deal with ambiguity, usually in game development we have the foresight to constrain problems (particularly frequent problems) which allows for us to pursue optimizations specific to our problem. The last approach I want to discuss has a nice optimization when we know that the shape is specifically a rectangle, and it uses this extra piece of information to achieve our most efficient solution.

Here is the projection inequality (introduced in an earlier chapter), with point M and side AB:

$$0 \le (AB \cdot AM) \le (AB \cdot AB)$$

This constraint is checking if M, projected onto AB, falls between A and B. If A and B form the rectangle's "top" edge, then this checks if M is within the left and right bounds of the rectangle. This check solves one dimension of the problem, but we also need to check if M is within the top and bottom bounds of the rectangle.

The second check is achieved in the same way, we just need to pick an adjacent side of the rectangle to the first side we checked. In this case, I will use BC.

$$0 \le (BC \cdot BM) \le (BC \cdot BC)$$

If both of the inequalities hold, then M is within the rectangle. As reference in an earlier chapter, you can watch a worked example of this equation on my YouTube channel in the video titled "Applications of the dot product inequality" [1]. Please note that the reason that we only needed to check two adjacent sides is because each represented a symmetric pair. Together both pairs accounted for all four sides of the rectangle. If the shape was a hexagon, we would instead need to check three adjacent sides.

Practice Problem 11.2: Is point M within a convex polyhedron?

Solution: Earlier we explored an approach (to check if a point was contained in a polygon) that considered each edge of a polygon to check if a point was on the correct side of each edge. There is a similar approach to this problem, but instead of checking each of a polygon's edges, we will check each of the polyhedron's faces. If a point is on the correct side of every face's plane, then the point lies within the polyhedron. As covered in a practice problem of the previous chapter, we can determine if a point is below a plane using a vector from the point to the plane and the plane's normal vector. If a point is below all of a polyhedron's faces, then it lies within the polyhedron. Figure 11.4 visualizes this approach by showing the normals of three of six faces in a rectangular prism.

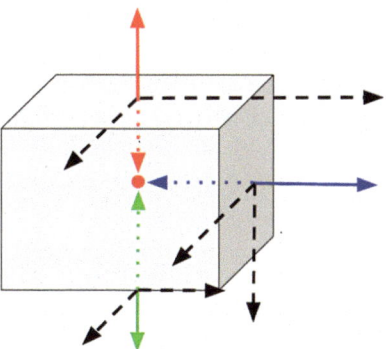

FIGURE 11.4 To check if the point is in the polyhedron, check if it is below each face's plane.

Similar to the approach with a rectangle, a rectangular prism has symmetric sides. So, while you may need to check every face for a generic polyhedron, for a rectangular prism you only need to check three faces which share the same vertex.

SIMPLIFYING COLLISION DETECTION

Because collision checks are so frequent and expensive in real-time games, there are many tricks that developers employ to reduce and optimize collision detection. A large part of the work happens while designing the data. Rather than model a 3D object's bounds exactly, developers will simplify its collision shape. As we explore several approaches consider their tradeoffs with respect to collision fidelity and runtime performance. There's no "right" answer here, and many games use a combination of approaches based on the needs of different objects.

CIRCLES

The first collision "trick" we will cover to simplify a polygon's collision bounds, is to circumscribe it, which means "draw a circle around it". Circles are, frankly, amazing. It's hard to describe their beauty in just words. There is something very special about how they are somewhat defined implicitly. Consider, in contrast, that a triangle takes three points to define, a square takes four, and so on. But amazingly, when you get to an infinite number of sides, you only need one point and a radius to implicitly define them all. The circle is, on one hand, conceptually something like a very complicated infinity-sided polygon, but on the other hand in practice it is the simplest shape we can work with.

A consistent theme through this book has been exploration of 'generalized' problems where we might solve a triangle collision problem by first solving a generalized version of the problem that considers the triangle's plane instead. In that vein, circles are our 'generalized' polygons (though by strict maths rules, circles are not technically polygons). Once we can determine that a collision intersects the circular bounds of a circumscribed polygon, we can consider the collision colliding or we can, like in the triangle problem, increase the specificity of the problem to test the collision against the inner polygon. This process of progressive specificity is called "phasing" and a game's collision detection algorithm might complete several phases of optimization in this manner. We will explore these phases starting with circle-to-circle collision.

Here is the formula for a circle:

$$\left(x - h\right)^2 + \left(y - k\right)^2 = r^2$$

- x and y are input coordinates.
- h and k are scalars that represent the circle's offset from the 2D origin.
- r is a scalar radius.

Though useful to know, I doubt you'll see this formula pop up in actual development. In practice, circles are usually just defined by a point at their center and a scalar radius. This assumes that we know what plane a circle lies in which is normally the case, if not then we also need a normal vector to know how the circle is oriented.

Practice Problem 11.3: Check if two circles A and B are overlapping. Assume these circles are in the same plane.

Solution: Now we can see the true power of the circle. If A and B are in the same plane, then we just need to check if the distance between them is greater than the sum of their radii. Anything closer to a circle than the circle's radius will be in the circle. To find the distance between two points we need to use the distance formula which requires a square root. An extremely common trick in this case is to not take the square root and instead compare the squared distance to the squared sum of the radii. This approach is also used for sphere collision tests.

As an example for this approach, consider two circles at (0,0) and (0,1). These points are only one unit away. So if one of the circles has a radius greater than or equal to one, we know with certainty that these circles are colliding.

SPHERES

I mentioned earlier how it was somewhat fascinating that a circle needed only a point and a radius to define itself. For spheres this fascination is only further amplified. They add an entire additional dimension to circles yet actually simplify the data necessary to define them. While you need the plane of a circle to define it in 3D space, spheres have no such requirement. They truly need only a center point and radius. Due to this incredibly small space requirement and for their extremely fast collision detection, spheres are used all the time in game development.

Here are some formulae to know for the sphere:

$$(x-a)^2 + (y-b)^2 + (z-c)^2 = r^2$$

- Scalars x, y, and z represent input coordinates.
- Scalars a, b, and c represent the offset of the sphere from the origin.
- Scalar r is the radius.

This formula is very much like the circle formula except with an extra dimension. It's rarely used directly, because in code spheres are usually just represented with a point and a scalar radius.

$$\| X - C \|^2 = r^2$$

This is the same formula as the previous formula except the coordinate components have been combined into two points. X is the input point, C is the circle's center, and r is still the scalar radius. The border pipes "||" represent that we are taking the L-2 norm of the resulting subtraction of X–C, aka the magnitude of CX.

$$(P-C) \cdot (P-C) = r^2$$

Of course, any squared magnitude can instead be represented by a dot product.

Surface area : $4\pi r^2$

For comparison, a circle's surface area is (πr^2), the sphere is therefore four times the surface area of a circle with equivalent radius.

Volume : $(4\pi r^3)/3$

This volume is derived by splitting a sphere into many pyramids, where the points of each pyramid meet in the sphere center. The volume of a pyramid (1/3 * base * height).

The height of each pyramid is r, and the sum of all bases is the aforementioned surface area of $4 \pi r^2$. I highly recommend the YouTube video "Understanding the Volume of a Sphere Formula [Using High School Geometry]" by mathematicsonline which includes a visual derivation of the formula using this process [2].

SPHERE COLLISIONS

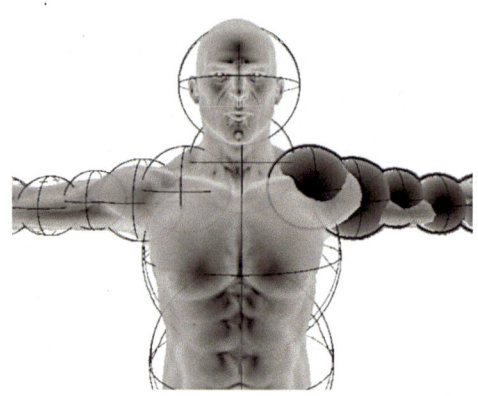

FIGURE 11.5   Spheres placed on a model to simplify collision.

Figure 11.5 illustrates collision spheres, sometimes called "hit spheres", placed on a humanoid model. These spheres usually have no meshes, but I have added some on the right side of the figure to better illustrate their positions. Rather than checking if something collides with the actual humanoid model, we can check its hit spheres. This is an efficient way to represent collision data for complex polyhedrons. Hit spheres often included additional markup to support things like the head hit sphere taking extra damage.

Practice Problem 11.4: Determine if two spheres A and B intersect.

Solution: This problem can be solved just like the circle-to-circle intersection. Simply test if the distance between the center of both spheres is less than the sum of their radii. For performance, don't find the true distance, just compare the squared distance against the squared radii sum.

Practice Problem 11.5: Determine if a ray, defined by point R and normalized direction vector $\hat{v}$, intersects a sphere, defined by point S and radius r.

Solution: Since we only need to determine if they intersect, we can simplify this problem to instead determine how close the ray gets to S. If the ray gets within r units of S, we know that there is an intersection.

Our first step is to project the sphere's center onto the ray's line, as shown in Figure 11.6. I have labeled the projected point as E. Now we can calculate vector **w**, which is E − S. If **w**'s magnitude is greater than the radius, then there is no intersection.

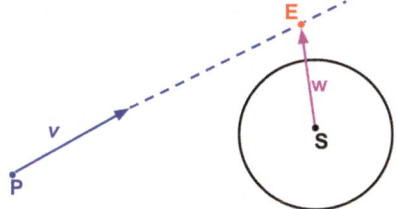

FIGURE 11.6   A sphere's center point S is projected onto the line defined by vector $\hat{v}$.

If there is an intersection with the line defined by $\hat{v}$, there is not necessarily an intersection with the ray defined by $\hat{v}$. Rays extend an infinite distance but only in one direction. We need to make sure E is in the direction of the ray and not behind the ray. You can use the dot product to confirm that PE is in the direction of $\hat{v}$.

If the projection of the sphere's center is within the sphere radius and on the ray then the intersection is valid.

Practice Problem 11.6: Determine the exact intersection between a ray, defined by point R and normalized direction vector $\hat{v}$, and a separate sphere, defined by center point S and radius r. Your answer should be one of three possible outcomes: zero, one, or two points of intersection.

Solution: For this solution I think the most intuitive approach is to simply plug in the ray's parametric formula into the sphere's formula and solve for when the sphere equation is satisfied.

Here are the basic formulas:

$$\text{Sphere}: (X - S) \cdot (X - S) = r^2$$

$$\text{Ray}: R + t\,\hat{v}$$

We can then plug the ray formula into the sphere formula as input X.

$$\left(R + t\hat{v} - S\right) \cdot \left(R + t\hat{v} - S\right) = r^2$$

Time to rearrange. We can now see the term "R−S" emerge. This is the vector from the sphere's center to the ray's start as shown in this diagram:

$$\left(R - S + t\hat{v}\right) \cdot \left(R - S + t\hat{v}\right) = r^2$$

Because "R−S" is known and in order to simplify the problem, we will replace the term with a new variable, **d** which we can easily calculate by subtracting point S from R.

$$\text{Let}: \mathbf{d} = R - S$$

$$\left(\mathbf{d} + t\hat{v}\right) \cdot \left(\mathbf{d} + t\hat{v}\right) = r^2$$

Expand the dot product.

$$\left(\mathbf{d} \cdot \mathbf{d}\right) + \left(\mathbf{d} \cdot t\hat{v}\right) + \left(\mathbf{d} \cdot t\hat{v}\right) + t^2\left(\hat{v} \cdot \hat{v}\right) = r^2$$

Since $\hat{v}$ is a unit vector, its magnitude is one and we can simplify the "$\hat{v} \cdot \hat{v}$" operand out of the equation.

$$\left(\hat{v} \cdot \hat{v}\right) = \|\hat{v}\|^2 = 1^2 = 1$$
$$\left(\mathbf{d} \cdot \mathbf{d}\right) + \left(\mathbf{d} \cdot t\hat{v}\right) + \left(\mathbf{d} \cdot t\hat{v}\right) + t^2 = r^2$$

Simplify. And now does this form look familiar? It's quadratic! We can now use the quadratic formula to solve for our only unknown: t. The quadratic formula's ABC inputs are included below.

$$\left(\mathbf{d} \cdot \mathbf{d}\right) + 2\left(\mathbf{d} \cdot t\hat{v}\right) + t^2 - r^2 = 0$$
$$A = 1$$
$$B = 2\left(\mathbf{d} \cdot \hat{v}\right)$$
$$C = \left(\mathbf{d} \cdot \mathbf{d}\right) - r^2$$

Our solutions are:

$$t_1 = -B + \text{rad}\left(B^2 - C\right)$$

$$t_2 = -B - \text{rad}\left(B^2 - C\right)$$

The term under the radical, aka the discriminant, determines how many intersections are valid.

If discriminant < 0, there are zero intersection.

If discriminant $== 0$, there is one intersection.

If discriminant > 0, there are two intersections.

I highly recommend Christer Ericson's Real-time Collision Detection, sometimes called "the orange book". It includes this derivation on page 177, "5.3.2 Intersecting Ray or Segment Against Sphere," in addition to many more practical problems in collision detection [3].

Practice Problem 11.7: Determine the exact intersection between a plane, defined by point P and normalized normal vector $\hat{v}$, and a sphere defined by center point S and radius r. Your answer should be one of three possible outcomes: zero intersection, a point, or a circle.

Hint: The shortest path between two structures is always going to be along a straight line between them. For this problem, we just need to compare the straight line, between S and the plane, with the radius of the sphere. How can we determine this line?

Solution: This is another classic "find the right triangle problem". In this case we already have two points of interest, S and P. We just need to know if there is a point of intersection, which I'll call "I". Projecting S onto the plane will provide us with a straight line between the plane and sphere, from S to I.

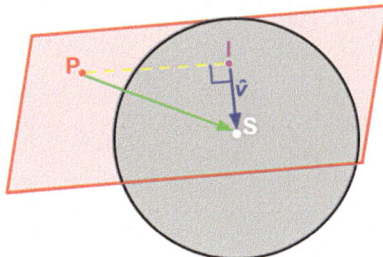

FIGURE 11.7 A sphere's center point S is projected onto the plane, along its normal $\hat{v}$.

Point I forms the third vertex of the right triangle ISP, which completely solves the problem.

- If ||IS|| > sphere radius, there is no intersection.
- If ||IS|| = sphere radius, there is one point of intersection. The point is I.
- If ||IS|| < sphere radius, there is a circle of intersection. The circle is centered at I, oriented in the plane, and with a radius that is the length of ||PI||.

Once you have three points of a triangle, you know everything about it, so the projection is the only thing we need to do math-wise. And we covered projecting a point onto a plane in a previous chapter.

BROAD-PHASE COLLISION DETECTION

Similar to how a circle can be used to circumscribe any polygon, a sufficiently large sphere can contain any polyhedron. Because sphere collision tests are so efficient, developers can wrap complex polyhedrons in spheres and compare if these outer spheres collide to know if it is even possible for the inner polyhedrons to collide. This "broad phase" can save significant time by culling polyhedrons from the more expensive "narrow phase". There are many ways to wrap complicated shapes into simpler shapers for performance. While bounding spheres are extremely fast, they are not an efficient bounding volume for tall or long objects.

Oriented Bounding Boxes

Oriented bounding boxes (OBBs) are an alternative bounding volume type. Their use is pretty self-explanatory, we simply take a rectangular prism, and set its width, height, and length to encapsulate a more complex polyhedron. As the "oriented" term in their name implies, OBBs can be rotated to best fit the objects they are bounding.

Practice Problem 11.8: Determine if two OBBs, A and B, are colliding.

Solution: One of the most efficient ways to solve this problem is the Separating Axis Theorem. The principle behind this approach is to project all of the vertices from two convex shapes onto a series of axes and determine if those points define an overlap. It can be used for any convex polygon or polyhedron.

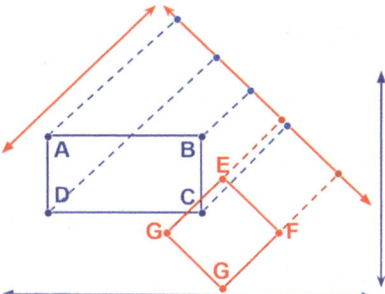

FIGURE 11.8 SAT projects every point of both structures onto every axis of both structures.

Every point must be projected onto every axis of the two structures. Since OBBs are rectangular prisms, their faces come in pairs that share axes. This reduces the number of total axes we need to project to, and results in OBBs being very efficient compared to more complex polyhedrons. With the points projected onto an axis, we can use the dot product inequality to determine if any of one structure's points fall within the other structure's points. While Figure 11.8 illustrates a 2D example, the same process is used in 3D except instead of each edge acting as an axis, the plane of each face acts as an axis.

You can view a worked example of this approach via StackExchange [1].

Axis-Aligned Bounding Boxes

Axis-Aligned Bounding Boxes (AABBs) are like OBBs except that they cannot be oriented. Instead, they are always locked to the world axes which provides faster collision detection.

Practice Problem 11.9: Determine if a ray defined by point P and vector **v**, intersects with an AABB.

Solution: This is another problem where we want to solve the generalized version first, then constrain the solution by the specifics of the problem. In this case we recognize that a ray is a line, with restrictions, and each side of an AABB is a plane, with restrictions. An AABB has six planes in total, but they are symmetric and can be thought of as three parallel pairs. Each plane pair represents the max and min of the AABB in a certain dimension. For example, Figure 11.9 illustrates that the positions where **v** intersects the AABB represent the min or max of the AABB with respect to one of the xyz axes. One approach to solve this problem is to determine the positions of intersection between **v** and each of the AABB's generalized planes, then determine which ones occur within the bounds of this specific problem.

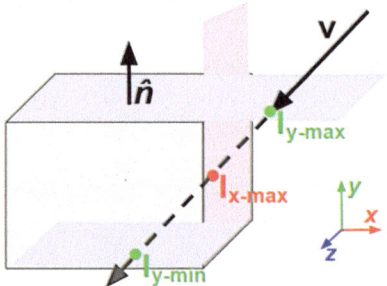

FIGURE 11.9   Vector **v** shown intersecting the AABB at the min and max of its y-axis.

First, we determine the t-min and t-max for each of the three axes, resulting in a total of six values: $t_{x\text{-min}}$, $t_{x\text{-max}}$, $t_{y\text{-min}}$, $t_{y\text{-max}}$, $t_{z\text{-min}}$, and $t_{z\text{-max}}$. Calculating a particular t value is simply the subproblem of determining when a ray intersects a plane which we covered previously. However, we can simplify the subproblem since we know these planes are not just any planes but specifically planes defined by the main xyz axes. So solving $t_{y\text{-max}}$ is actually just:

$$t_{y-\max} = \left(\text{AABB}_{y-\max} - P.y\right)/V.y$$

This expression above is simply determining how far from the vector's starting point, P, that the **v** would need to travel to reach the maximum y value of the AABB. Then we divide that distance by **v**'s magnitude in that direction to determine how many instances of the vector **v** we would need to line up, end-to-end, in order to traverse that distance. We've essentially simplified the subproblem down into just a 1D problem, thinking exclusively about y values. This simplification is permitted due to an AABB's restricted orientation and is its key benefit.

Some t values will be negative, indicating that while they're on the line defined by $\mathbf{v}$, they won't be on the ray defined by $\mathbf{v}$. You might also discover that a component of $\mathbf{v}$ is zero in which case the vector is parallel to a plane. In these cases, the y min and max values are usually set to sentinel values of negative infinity and positive infinity respectively. Once we have a t value for each bounding plane, we can convert them to positions. Here's an example:

$$I_{y-max} = P + \mathbf{v}\, t_{y-max}.$$

With our intersection points determined, we have completed the generalized version of the problem. Now we can add the problem's constraints back in to solve this specific problem. This nuance is demonstrated in Figure 11.9 where the vector is shown intersecting the AABB's y-max plane before it reaches the actual AABB. The solution max value will be the closest of our I-max points to the origin. The solution min value will be the furthest I-max value from the origin. We also need to confirm that points are on the ray rather than the line defined by $\mathbf{v}$. You can view a worked solution of this problem in the YouTube video "WebGL 2.0 : 046 : Ray intersects Bounding Box (AABB)" by SketchpunkLabs [5].

Practice Problem 11.10: Determine the overlap of two AABBs, A and B.

Solution: Here we see the true power of AABBs. Since they are both aligned to the same axes, and those axes are world axes directly corresponding to their coordinates, we simply need to compare their min and max values along each axis to determine an intersection. After all the work we've done to explore some complicated intersections, the simplicity of AABBs feels like a breath of fresh air. Here's the solution in just a few lines of pseudo code for a 2D example.

First, we look at each axis to determine if the AABBs overlap at all. For example, if the minimum x value of A is further along the x-axis than the maximum x value of B, then A itself is completely further along the x-axis than B.

```
noOverlap = false
noOverlap |= (A_x-min > B_x-max or B_x-min > A_x-max)
noOverlap |= (A_y-min > B_y-max or B_y-min > A_y-max)
if(noOverlap){ return 0 }
```

If there is an intersection, it will be defined by the inner max and min of the two structures. This can be determined by finding the minimum of the max values, and the maximum of the min values.

```
widthOverlap = min(A_x-max, B_x-max) - max(A_x-min, B_x-min)
heightOverlap = min(A_y-max, B_y-max) - max(A_y-min, B_y-min)
```

If the overlapping width or height are negative, that means there is no overlap. This would have been caught in the earlier noOverlap check. You can view a worked solution of this approach in the YouTube video "Math for Game Developers - Trigger Areas (AABB Intersection)" by Jorge Rodriguez [6].

Spatial Partitioning

AABBs, like spheres discussed earlier, can be used to enclose complex polyhedrons to optimize broad phase collision detection. But AABBs can also be used to enclose other AABBs. It's very common for developers to partition their world into a hierarchy of AABBs to simplify the number of collisions they need to process. For example, if a game world has four quadrants, and the only objects of the world are in separate quadrants, then we know that none of the objects are intersecting.

There are many types of trees that can be used to store the data of hierarchical AABBs. This includes trees that contain nodes of four children called quadtrees (common in 2D games), and trees that contain nodes of eight children called octrees (common in 3D games).

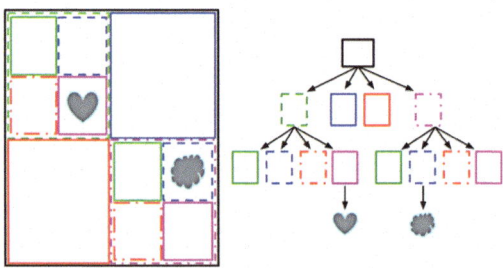

FIGURE 11.10   A game world is partitioned into a hierarchy of AABBs, organized in a quadtree.

WORKS CITED

[1] M. Ventures, "Applications of the dot product inequality," *YouTube*, Aug. 16, 2024. [Online]. Available: https://youtu.be/srqhPL4nWlI?si=BKCus1YN8SCShNBY

[2] Mathematicsonline, "Understanding the Volume of a Sphere Formula [Using High School Geometry]," *YouTube*, Aug. 6, 2023. [Online]. Available: https://youtu.be/xuPl_8o_j7k. [Accessed: Aug. 22, 2024].

[3] C. Ericson, *Real-time Collision Detection*. Natick, MA: CRC Press, 2005, p. 177.

[4] J. Shbrntt, "OBB vs OBB Collision Detection," *Game Development Stack Exchange*, https://gamedev.stackexchange.com/questions/25397/obb-vs-obb-collision-detection. Accessed Aug. 24, 2024.

[5] Sketchpunk Labs, "Ray vs. AABB Intersection Explained," *YouTube*, Oct. 22, 2023. [Online]. Available: https://www.youtube.com/watch?v=4h-jlOBsndU&t=2120s. [Accessed: Aug. 22, 2024].

[6] J. Rodriguez, "Math for Game Developers - Trigger Areas (AABB Intersection)," *YouTube*, [Online]. Available: https://www.youtube.com/watch?v=ENuk9HgeTiI. [Accessed: Aug. 25, 2024].

PART III

Additional Questions

Behavioral Questions

12

As you gain responsibility through your career, your interviews will begin to focus more on how you approach problem solving. These questions ask candidates to think "big picture" and consider not just how to solve a specific technical problem, but how to effectively collaborate across different teams and lead larger projects. An example of this question is "tell me about a time when you had to get a project or initiative completed with limited resources."

PREPARING STAR ANSWERS

The best way to prepare for behavioral interviews is to write a short list of the technical challenges you have overcome and to format stories about those challenges into the STAR format.

The star system breaks stories into segments: Situation, Task, Action, and Result. I also advise my students to always consider Legacy, Metrics, and Regrets as well. Here's a list of each segment and questions to help you cover them.

- **Situation**: What was the inciting action of your story? Did a user find a bug? Did the creative director request a feature?
- **Task**: What was your plan to address the situation? How did you know your plan was the right direction?
- **Action**: To what extent did you follow through on the plan? Did you run into any surprises?
- **Results**: How was the situation resolved? What were the results?
- **Legacy**: How did addressing this situation affect future situations? Did your actions help establish any norms or processes?
- **Metrics**: What quantitative data can you share about the situation? How did you measure your results?
- **Regrets**: How would you have changed your approach with the benefit of hindsight?

DOI: 10.1201/9781003565550-15

Not every story is going to have great metrics, or an interesting distinction between Task and Action, but breaking the story up into these segments can help you think through the core details. It can also help you compare potential stories you may want to share. For example, if one story has a great quantifiable result, it may be more impactful than a story where the results of your work were hard to discern.

STAR Example

Situation: While I was working on Call of Duty Warzone (WZ), a battle-royale FPS, the team wanted to have a marketing beat for the winter holidays, featuring themed in-game content.

Task: The original design from the creative director was to create Christmas trees that randomly spawned into the map with loot boxes, represented as Christmas presents, beneath them. I saw this as an opportunity to fix a long-standing issue that I identified players had in our game. Unlike other battle royales, WZ had uniformly distributed loot. Other games had 'hot zones' known for high concentrations of PVP combat and great loot. These hot zones provided players with a choice of whether they wanted to grab the best loot, at the risk of facing more enemies, or play it safe by starting their game in colder zones of the map. Players had long complained about the sameness of our map's loot distribution, preventing this interesting decision. I saw the Christmas content as an opportunity to fix this problem in our core gameplay while delivering the marketing beat required.

Action: I proposed to the creative director that we could make these trees have some of the best loot in the game and we could require players to defend the trees in order to earn that loot. By requiring players to stand near the tree, we could broadcast their position, causing nearby players to be alerted of their location and increase the likelihood of conflict in those high-loot areas. The design was approved, and I started coding. During development I created a runtime tool that allowed my designer to walk through the game world and test placements for the different potential tree locations. Normally testing a tree placement required placing it in the world editor and completing a lengthy map build. But by placing it within the runtime environment, the designer could instantly view it from multiple angles to consider if it would be a good focal point for combat.

Results: The feature was very well received by the team and the WZ player base. The hectic fray surrounding the trees was a nice change of pace for most players and many streamers created dedicated videos on the gameplay including tutorials and highlight reels.

Legacy: I intentionally designed versatile code that I thought could be used in future events to provide similar gameplay. Following the winter holidays, I suggested to the team that we use this mechanic for a St. Patrick's Day event with a pot of gold in the center instead of a tree. Unfortunately, I was not successful in this endeavor because leadership was concerned that if we created a similar event, the players would think we are lazy. I contested that some content was better than no content but ultimately was not successful in securing approval for this work. While my efforts to gain support for additional holiday content were unsuccessful, the code proved useful later as it allowed us to quickly build events that required players stay in the vicinity of a checkpoint to complete an objective.

Metrics: I don't have hard data on the number of players that engaged with this mechanic, I wish I did! It would have been smart of me to record this data and identify quantitative metrics to measure if it was effective at incentivizing more varied play.

Regrets: One specific tree spawn point was placed in an area that was not known for high loot but that did have high player traffic. Several players said that the audio cues for this particular tree were somewhat annoying when they were trying to focus on fighting in that region. This was mostly due to an underground area near the tree, where players could hear the tree despite it not being relevant to them. With the benefit of hindsight, I wish that I had specifically worked with my level designer to ensure that trees were only placed in otherwise low-traffic zones and zones without underground areas nearby. With additional time, I could have also developed a system, to not propagate sounds into the ground below the tree, to replace the simple radius-based check that already existed.

COMMON STAR QUESTIONS

Here are a few common questions that can help you think of stories to tell in the STAR format. Some stories can be used across a variety of STAR questions, though generally you want to share different stories within different interviews at the same company to showcase your many skills. Ideally, you should prepare at least four stories that cover the breadth of these questions.

Tell me about a time when you…

- Learned from a bad decision.
- Leveraged the strengths of a teammate.
- Challenged the status quo.
- Took a big risk.
- Used data to inform your decision making.
- Balanced player expectations with your development team's objectives.
- Went beyond your normal duties to help your teammate.
- Completed an objective with limited resources.
- Used a metric to identify a need.
- Disagreed with your leadership.

- Dug deep, through several layers, to solve an underlying problem.
- Realized you were pursuing the wrong goal.
- Had impact beyond your immediate team.
- Worked outside of your comfort zone.
- Made a mistake.

Asking Questions

13

EASY QUESTIONS

Game development studios can vary a lot with respect to how they make their games. At the end of the interview, you have the opportunity to ask them questions about how their studio works. In theory, this time is presumably for a candidate to identify any red flags that would make them not want to work at the studio. In practice, it's probably better to do your research on a company before the interview stage because searching for those red flags can unfortunately sometimes hurt your candidacy. I never want any of my students to take on a role that they're not going to be happy in. But, strictly focused on having you pass the interview, I would encourage you to focus on this section of the interview as an opportunity to show your interest and passion in their company by asking positive questions. For example, this is a great question:

> I watched your team's GDC talk on the AI token system. I thought it was really cool, and tried to implement some of the approaches you used into a little game that I am making. However, one of the issues I ran into was that players could very easily discern when an AI didn't have an attack token and was instead just standing around, seemingly doing nothing, waiting for their turn. What are some of the ways your team has explored dealing with this issue?

An easy question like this shows them you're interested in the company, you're familiar with their work, and you want to learn more about the problem solving they do on a day-to-day basis. If the company has something like a GDC talk that is a great way to dig into some of the problems they have worked on. Or you could read a review for their latest game to brainstorm what kinds of problems the team has recently focused on from a player's perspective.

The idea behind these easy "softball" questions is mainly just to prop you up as a candidate. In my experience interviewing people at competitive studios, we have turned candidates away because we thought they may have been talented but not genuinely interested in our work. For example, I was shocked at how many people that applied for

---

DOI: 10.1201/9781003565550-16

a job at a prominent first-person-shooter developer, did not actually enjoy first-person-shooters. That is fine for some roles, but not gameplay engineering.

My favorite question of this nature is: "If you were to hire me with a start date in one month, what should I do within the month to prepare?" This is a sly question because it asks them to think about you as someone who has already been offered the job. You're playing with some psychology here, painting yourself in their mind with a positive perspective.

HARD QUESTIONS

Maybe there was not enough information online to adequately research a company. And you were not able to schedule informational interviews with the team prior to a formal interview circuit. In these cases, you may want to ask questions that are a bit harder, and at times may touch upon negative subjects. Sometimes landing the right job, is more important than landing just any job. It depends on your needs as a candidate. But most of these questions can be asked in a non-confrontational manner, even if they are a bit harder for interviewers to answer.

The first topic candidates usually ask about is crunch. The game development industry has a horrible history of mistreating its developers, please study the history of events like EA Spouse. The main things to ask about crunch are "what's the official policy?" and "what's the actual in-practice reality?" Here are some example questions:

- When was the last time your studio crunched?
- When your studio last crunched, what did that mean? Did people work 10-hour days? Or 6-day weeks? Was it optional? Was it optional but like 99% of people did it, so it really wasn't optional but that is just what management said when people asked? etc.

Team composition is also an important topic for many candidates to ask about. If you are a woman and feel that you may be uncomfortable joining a team with no other women, then you probably want to know what the team you are working with is going to look like from that perspective. Probably the most important factor on a team is its size. A team of four operates way differently than a team of forty, there are positives and negatives to both.

- What is the size of the studio?
- How are teams divided within the studio?
- What percent of the work force is engineers?
- How many of the engineers are woman, people of color, etc.?
- What percent of the team works from home, hybrid, in-office?
- Is there a dedicated tools and AI team?

Here are some questions to explore the development practices of a team. I'd be particularly wary of a team's creative control. It's very common nowadays for a single studio to lead creative development of a game and hire outsourced studios to do their drudgery.

- How often does the development team participate in testing?
- Does the design team do any scripting? Does the engineering team do any scripting?
- How often does the studio have a studio-wide meeting? What about the company division or company as a whole?
- What is the state of internal development tooling, how frequently does something like the nightly build fail?
- What is the role of production on your team? Do producers lead creative decision-making? Do you use SCRUM or track velocities?
- How often are reflections / post-mortems conducted?
- Does this team have sole creative authority over the work, or are we a satellite studio subject to the whims of a different studio?
- Is the creative leadership who started this project located at this studio? Do they still work on this project?

Sometimes it's helpful to just throw out a generic STAR behavioral question to explore a team's dynamics. For these questions it is helpful to be specific.

- How have the art team and the design team previously resolved differences of opinion with the engineering team?
- What is a recent argument your team had? How was it resolved?

PART IV

Practice Test

Practice Test

14

TEST-TAKING TIPS

If you ever feel stuck on an interview question, it is important to remain calm. Try and ask some clarifying questions to first determine if you can re-contextualize the question into your knowledge space. Here are some examples:

- I'm not familiar with that term, I think it may just be that I have not heard it described or pronounced in that manner, could you please rephrase the question?
- I haven't directly approached that specific topic in my work thus far, but I think I may be able to answer this question. Could you contextualize the problem in an example for me?
- I'm not sure how I would start on such an abstract problem, could you provide a grounded starting point for me to work off of?

If those questions can't deliver you to a productive starting point, you should probably just ask for help. Your goal at that point is to demonstrate that while you may not know the specific knowledge they were testing for, you are intelligent and can express your reasoning skills in addition to showcasing any adjacent subject knowledge you may have.

- I haven't worked with that specifically. I'm familiar with X which I think could be applicable to the question you asked me, but I think X is only relevant to the context of Y. Could you explain to me a bit more about how the system you are describing works?
- I think that in pursuit of approaching this problem I would first refresh my knowledge with respect to X technology. Could you please provide me with more information about X technology that might be useful for this problem?

DOI: 10.1201/9781003565550-18

Lastly, even if you completely fail the question, always demonstrate an appetite for learning! It sucks as an interviewer when you ask your interviewee an interesting question and they flatly respond that they don't know and clearly indicate that they just want to skip the question. As a candidate, make sure to communicate that you have a growth mindset and that you are excited to learn about things you don't yet know about!

- That's not something I've worked with before. I think I would need to do some research to provide you with a satisfactory answer. Could you provide me the solution you were looking for? I definitely want to follow up and self-study this when I get home tonight.

PRACTICE PROBLEMS REVIEW

Let's take a walk through some of the previous chapters to revisit some of the topics we covered.

Fundamentals

Practice Problem 14.1: What's the fastest way to tell if 2 signed floats have the same polarity?

Hint: Floats store their polarity (whether they are positive or negative) in their sign bit. It's their leftmost bit.

Solution: Compare the sign bit of both numbers to see if they are equal. But how would we do this comparison, might an AND or XOR work?

Let's walk through our bitwise operators to see if any of them can help us here... What if we try AND? If you AND two numbers then the resulting value's ith bit will be 1 if the ith bits of both input numbers was 1. For example, 1001 AND 1100 = 1000. So, if we AND our numbers then we can just check the first bit of the result to see if it is a 1. It will be a 1 if both of the input values were 1. But this approach wouldn't quite work to check if the two input values were both 0. That's because if both of the input values had a 0 as their first bit, the resulting value's first bit would be 0 which would be indistinguishable from a 0 we would get if the two input values had different values as their first bit. Put a bit more simply... since 0 & 0 = 0 and 0 & 1 = 0. We can't tell which case the 0 came from.

What about if we use XOR? If you XOR two numbers, then the resulting value's ith bit will be 1 if the ith bits of both input numbers are not the same value. For example, 1001 XOR 1100 = 0101. So, if we XOR our numbers then we can just check the first bit of the result to see if it is a 1. It will be a 1 if the input values had different values in their ith bit. This is exactly what we need! If the resulting value of a XOR has a 1 in the first bit, then that means the two values had different values in their ith bit meaning they are of different polarity.

So, the solution is to XOR the two values, then check the first bit to see if it is 0 which would mean that the two input values had the same sign bit. Try it out with some examples to test.

Practice Problem 14.2: Below is a simple nested for-loop that traverses every element of a 2D matrix, we could choose option A or B to perform the desired "++" operation. Which is the faster option and why?

```
for (int X = 0; X < MAX; X++) {
    for (int Y = 0; Y < MAX; Y++) {
        // Option A:  matrix[ X ][ Y ]++;
        // Option B:  matrix[ Y ][ X ]++;
    }
}
```

Hint: Consider the cache locality of the data.

Solution: In this case, A is the faster option. It's faster because it uses consecutive values within the matrix. C and C++ store matrix values in row major order so that means when we grab a matrix index from memory it is going to come with a cache line comprised of the other values in its row in addition to adjacent rows. We can reduce the frequency of cache misses by primarily iterating through the rows of data rather than by primarily iterating through the columns which would have the processor jumping around memory and missing the cache more frequently.

The question comes from the opening of Scott Meyer's talk "Cpu Caches and Why You Care" available on YouTube [1].

C++

Practice Problem 14.3: Explain the difference between "const pointers" and "pointers to const" in C++.

Hint: Remember that the placement of the const keyword can affect which aspect of the pointer and its value are constant.

Solution: A pointer to a constant value, such as "`const int* myPtr;`", indicates that the value being pointed to cannot be modified through this pointer, but the pointer itself can be changed to point to another address. A constant pointer to a value, such as "`int* const myPtr;`" indicates that the pointer itself cannot be changed to point to a different address, but the value at the pointed-to address can be modified. Of course, both of these can also be combined such as "`const int* const myPtr;`", to indicate that both the pointer and the value are constant.

Data Structures

Practice Problem 14.4: How would you implement a priority queue for managing game events with varying levels of urgency?

Hint: What data structure is designed around quick access for high "max" values?

Solution: A priority queue like this should be implemented as a max heap which provides immediate access to the event of maximum urgency. In this structure, events are stored with their urgency levels, and the event with the highest urgency (highest priority value) is always at the top. This allows efficient insertion and extraction of events based on their urgency.

Patterns

Practice Problem 14.5: You've been tasked with implementing an AI's attacks, which includes runtime decision making that considers world state. How would you compare the approaches of using FSMs versus behavior trees? Which do you think would be the best fit for this scenario?

Hint: The key difference between these patterns is that FSMs are stateful. How can that inform your decision?

Solution: FSMs excel at modeling discretized, stateful, behavior. If your enemy has a small number of attacks and those attacks control their logic for a duration of time, then FSMs are probably the way to go. If your AI's attacks are one-off actions that do not heavily depend on the AI's actions prior to the attack, then behavior trees may be a better choice. If the checks to perform an attack lend themselves to a hierarchical node-based design, then behavior trees will be easier to use for modeling this logic. Sometimes hybrid approaches are used where FSMs may include non-stateful states (sometimes called passthroughs or conduits) that will fall back to previous states if they do not present transitions to stateful states.

Multithreading

Practice Problem 14.6: Consider a scenario where a system has many threads sharing a resource that is protected by a single lock. When the lock is acquired by one thread, other threads are blocked and placed in a queue waiting for the lock to be released. As soon as the lock is released, the next thread in the queue acquires it, and the process continues. This scenario results in significant wait times for threads and frequent context switching, what alternative solutions can you propose?

Hint: Consider that some threads accessing the resource may not need to edit the data.

Solution: In this scenario, the issue is a "lock convoy", where multiple threads queue up for a single lock, causing high wait times and context switching. To address this, you can reduce lock granularity by using smaller, more targeted locks (consider lock striping), implement read-write locks to allow concurrent reads, or switch to lock-free data structures.

Maths - Vectors

Practice Problem 14.7: You are challenged to implement a player's backstab ability. The idea is that if a player can see an enemy, but the enemy cannot see them, then the player can deal a ton of damage with a single backstab attack. You have so far implemented the CanBackStab function, you must now implement the IsInFov helper function that will be used to test if the player can see the enemy and if the enemy can see the player. Assume an FOV spans 60°.

```
bool CanBackStab(const Transform& player, const Transform&
enemy) {
    const Vector3 playerToEnemy = enemy.position - player.
    position;
    const Vector3 enemyToPlayer = -playerToEnemy;
    const bool playerSeesEnemy = IsInFov(&player.forward,
    &playerToEnemy);
    const bool enemySeesPlayer = IsInFov(&enemy.forward,
    &enemyToPlayer);
    return playerSeesEnemy && enemySeesPlayer;
}

/* Determines if a line of sight falls within the field
of view (FOV) of an entity looking in the forward
direction: */
bool IsInFov(const Vector3& forward, const Vector3&
lineOfSight){
    // [Your code goes here]
}
```

Hint: Which vector operation can we use to determine the angle between two vectors?

Solution:

```
bool IsInFov(const Vector3& forward, const Vector3&
lineOfSight){
    const float kFovDegrees = 60.0f;
    const float kFovThresholdRadians = (kFovDegrees /
    2.0f) * (Math.PI / 180.0f);
    const float scalar = forward.magnitude * lineOf-
    Sight.magnitude;
    const float cosAngle = Math.DotProduct(forward,
    lineOfSight) / scalar;
    const float angle = Math.ArcCosine(cosAngle);
    return angle < kFovThreshold;
}
```

Practice Problem 14.8: You are challenged to implement the basic movement controls for a 3D fox game. After your first-pass implementation, you discover an issue. When you push the move joystick forward, the fox moves toward whatever direction it is currently facing, see Figure 14.1. This means that when the fox faces downward, the controller input to go up is down and the input to go down is up. This is confusing players who would prefer for up input to always move the fox away from them in the direction of the camera.

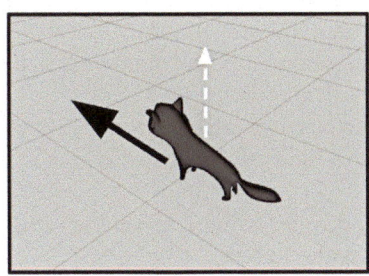

FIGURE 14.1 The gamepad forward input (black arrow on right) leads to the fox moving in its forward direction (black arrow on left) instead of in the camera's forward direction (white arrow) [2, 3].

It seems that this forward input is moving the fox with respect to its forward axis. You are challenged to make the movement interpret the player's input with respect to the camera's forward axis, regardless of the fox's orientation. You are given the following vectors as input:

- **j**: the vector of joystick input. It's a 2D vector and both components have values ranging from -1 to 1, indicating how much the joystick is pushed forward, back, left, and right. The first component has the x data (left vs right), and the second component has the y data (up vs down). In the case of the example, the input would be [0, 1], indicating that the joystick is fully pushed upward (y=1) and not at all pushed sideways (x=0).
- $c_{forward}$: the forward vector of the camera. It's a 3D vector and points directly into the scene.
- **u**: The fox's up vector. It's a 3D Vector. This vector always points up and implicitly defines the fox's walking plane.

Your output should be a 3D vector, **m**, which represents the direction the fox should move given the input parameters. Your answer should be in the format of pseudo code.

Hint: We want to move the fox in the direction of the camera's forward vector, $c_{forward}$, but we also need the movement vector to be within the fox's walking plane.

Solution:

- Normalize the joystick input as a vector, call it $\hat{\jmath}$.
- Project the camera's forward vector, $c_{forward}$, onto the avatar's walking plane, then normalize the resulting vector, call it $\hat{c}_{forward}$.
- Now you can scale vector $\hat{c}_{forward}$ by scalar $\hat{\jmath}_y$ to get the forward vector component of our final answer, aka m_y.
- For the sideways vector component of our final vector, aka m_x, we need the vector perpendicular to $\hat{c}_{forward}$.
- We can determine the vector perpendicular to $\hat{c}_{forward}$ by calculating its cross product with the fox's up vector, u. Let's call the normalized result $\hat{c}_{right}$.
- Now you can scale vector $\hat{c}_{right}$ by scalar $\hat{\jmath}_x$ to get the right vector component of our final answer, aka m_x.
- The final solution m is formed by adding together the partial solutions m_x and m_y.

Maths - Matrices

Practice Problem 14.9: Imagine we have two game objects, a dog and a skateboard, with the dog set up as a transform child of the skateboard. Each of these objects has its own rigid body transformation matrix, shown below.

Dog	Skateboard
$\begin{bmatrix} 1 & 0 & 0 & 10 \\ 0 & 1 & 0 & 20 \\ 0 & 0 & 1 & 30 \\ 0 & 0 & 0 & 1 \end{bmatrix}$	$\begin{bmatrix} 1 & 0 & 0 & 100 \\ 0 & 1 & 0 & 0 \\ 0 & 0 & 1 & 0 \\ 0 & 0 & 0 & 1 \end{bmatrix}$

Please use the matrix values above as the starting point for each of the three questions below, the questions do not build upon each other. Remember, the dog transform is a child of the skateboard transform.

#1 - What is the current local position and global position of the dog and the skateboard?

#2 - What would the resulting transform matrices look like if we set the dog's global position to $(150, 0, 0)$?

#3 - What would the resulting transform matrices look like if we moved the skateboard's local position to $(150, 0, 0)$? Assume that the dog will move with the skateboard.

Solution:

#1 - The matrices display local position in their right-hand column. The dog's is $(10, 20, 30)$, and the skateboard's is $(100, 0, 0)$. The skateboard has no parent, so its local position is its global position. The dog has the skateboard as its parent, so its local position is relative to its parent's global position. To get the dog's global position we can add its local position as an offset to the skateboard's global position: $(100, 0, 0) + (10, 20, 30) = (110, 20, 30)$.

#2 - If we set the dog's global position then the skateboard will be unaffected, so its transform matrix will not change. However, the dog's transform matrix will change. The right-hand column of $[10, 20, 30, 1]^T$ will become $[50, 0, 0, 1]^T$. The x value is 50 because the dog's local position is relative to the skateboard. So, starting from the skateboard's x value of 100, we need to add 50 to reach the final global position x value of 150.

#3 - The skateboard's new local position will be reflected directly in its transform matrix. The right-hand column of $[100, 0, 0, 1]^T$ will become $[150, 0, 0, 1]^T$. The dog will move as the skateboard moves but it will stay at the same position relative to the skateboard, so its transform matrix will not change.

Maths - Rotations

Practice Problem 14.11: Let's say we have two vectors, **a** and **b**. How can you form a quaternion to represent the rotation from **a** to **b**?

Solution: First, let's handle some edge cases.

If the vectors are parallel, which is equivalent to saying that their cross product results in the zero vector, then they are either facing the same way (positive dot product) or opposite directions (negative dot product). If they are facing the same way, then this is a zero rotation and the quaternion is $[1, 0, 0, 0]^T$ (the identity quaternion), where the vector entries correspond to w, x, y, z.

If they are facing opposite directions, then this is a 180° rotation. In this case we require an axis of rotation, which can be any vector perpendicular to **a**. Remember that the cross product will return a zero vector for parallel vectors so we can't cross **a** and **b** directly. Instead, we have a subproblem to find any vector perpendicular to **a** which we covered in the vectors chapter. Let's call the result $\mathbf{a}_\perp$.

If **a** and **b** are not colinear then getting $\mathbf{a}_\perp$ is as easy as taking the cross product between them.

Once you have $\mathbf{a}_\perp$, we need to normalize it to achieve our solution vector, $\hat{v}$. Then we can plug it into the formula below with the angle of rotation. In the edge case we identified the angle as 180°, but in the standard case we can just use the dot product to determine the angle.

Now that we have the perpendicular vector and the angle, we can simply plug those values into the quaternion formula for our solution:

$$\mathbf{q} = \left(w, \hat{v}\right) = \left[\cos\left(\theta/2\right), \hat{v}_x \sin\left(\theta/2\right), \hat{v}_y \sin\left(\theta/2\right), \hat{v}_z \sin\left(\theta/2\right)\right]$$

Maths - Planes

Practice Problem 14.11: Given a ray and a triangle, derive their intersection.

Solution: First recognize that this is a constrained version of the generalize problem to determine if a line intersects a plane. A ray is just a constrained line, and a triangle is just a constrained plane. Solve that easier subproblem first, covered previously, and if there is an intersection then take that point. We know the point is on the plane which contains the triangle, we just aren't yet sure if the point is in the triangle. So, all that remains is to check if the point is in the triangle, which is another subproblem we have already discussed. Also be sure to check that the point is on the ray and not just the line defined by the ray.

Maths - Polygons & Polyhedrons

Practice Problem 14.12: Determine if a ray, defined by point P and normalized vector $\hat{v}_{ray}$, intersects an OBB.

Solution: As usual, our best approach here is to simplify the problem into a general subproblem.

First let's deal with the ray. As we have done in pretty much every question of this nature with a ray, instead of addressing this specific problem directly, we will generalize the ray into a line. Then at the end of the problem, we can determine if the line's intersection point is within the valid range for the ray.

Second, we have the OBB to deal with. It's just a regular old rectangular prism but it has rotation. A neat trick which we will employ here is to nullify that rotation. Instead of treating both the line <u>and</u> the OBB as rotated, we can rotate the line to be in the local space of the OBB so that only the line will appear rotated. And if we can do that, then our OBB can be treated like an AABB and we can use the simpler approach, covered previously, for determining a ray's intersection with an AABB.

To transform the ray, we will apply the inverse of the OBB's rotation to both P and $\hat{v}_{ray}$. Remember that a rotation matrix is orthogonal, so its transpose is its inverse. If the OBB's rotation matrix is **R**. Then here is how we would find P_{local} and $\hat{v}_{ray\text{-}local}$:

$$P_{local} = \mathbf{R}^T \, P$$
$$\hat{v}_{ray\text{-}local} = \mathbf{R}^T \, \hat{v}_{ray}$$

Now using P_{local} and $\hat{v}_{ray\text{-}local}$ we can define a line that is local to the OBB and perform the previously covered ray-to-AABB intersection process.

Behavioral Questions

Practice Problem 14.13: Describe the process you would go through to address a bug reported by one of your game's players.

Solution: An anecdote in STAR format might be a good example to reference as you answer this question. Here are some high-level notes on what an answer might include:

First, discuss how bugs might be reported. This includes users manually reporting bugs via email or some in-app reporting tool. But it also includes analytics tools that track crash reports to catch issues that users might not report. If you're lucky, crash reports will come with a mini dump that you can open in an IDE to view state. But more than likely, you will only have a call stack or a list of reproduction steps.

Ideally, you'll be able reproduce the error. This is an important step because if you can reproduce it, you will be able to later determine if a potential solution truly fixed the issue. To help reproduce the issue, you might collaborate with quality assurance experts or talk to the user that reported the bug. If you're struggling to reproduce the bug, you might write temporary code and tooling to design an easier way to reproduce it. If you can't reproduce the bug, it's usually due to a lack of information about the problem. In these cases, you can still take action by improving your game reporting mechanisms so that additional data, such as print statements, will be provided if the issue happens again. Usually for these speculative fixes, we can add extra safety mechanisms like ensure statements or systems to govern memory management.

In order to help isolate the problem, you might use print statements, stack traces, and breakpoints (including data break points) to understand the state of the program at the time of the bug. It's often helpful to create runtime debug visualizations, using tools like IMGUI, to show how the code is interfacing with data under the hood. Additional tools might allow you to step frame-by-frame through a problem scenario or inject intentionally bad data into a system to force the bug.

Once you've addressed the bug, you'll want to try and prevent it in the future. Or improve the code so that if the issue does occur in the future, it will be easier to identify and resolve. You might incorporate asserts to detect bad states, but make sure that they are paired with mechanisms to resolve these states in release builds. It's a common mistake for developers to assume that since an assert or ensure fires in a debug build, then the bad state is addressed. That's not the case. You need to also implement a way to handle the problem scenario, just in case it happens outside of a development context. You might also add logs or tests that can help more easily detect the issue in the future.

Lastly, you may want to inform your teammates of the bug and how you fixed it. And if you're using a third-party engine like Unity or Unreal, you'll want to report any bugs that stem from the engine's code.

Practice Test

Before taking this practice test, imagine you are entering an interview for a specific game company you plan to apply to. Prepare any notes you think you may need to reference, including your resume. Then go through each of these questions as if they were a real interview. Try to answer the question without any external help. If you get stuck, read through some of the hints provided. And if those don't help, research for a few minutes online (Google, ChatGPT, etc.) before returning to the problem. Please keep in mind that a real interview will not take so long, it might include only one or two problems, but you might be expected to complete several interviews within a single day. Additionally, a real interviewer will guide you along, listening to your ideas and giving you feedback. Lastly, many of these problems have alternative approaches, so if you came up with a different way to solve the problem, that works too!

Practice Problem 14.14: Hello and welcome to [company you want to work at]. Have you played our recent game? What is a specific system that interested you in the game? How could it be improved?

Hint: Games can often be explained in terms of arcs and loops. Think about how your suggested change affects arcs and loops present in the existing design.

Solution: Make sure to focus on the positives. Your feedback should always be additive, rather than subtractive. For example: "ranged combat could be improved by adding X" versus "ranged combat should be removed". Certainly, avoid saying their game is bad. Even if you do think part of the game is bad or detracts from the overall experience, it would be more strategic to design your answer around an additive change that supplements the aspect you feel is lacking. It's likely that the person you are talking to worked on the game, perhaps even the bad parts.

Example Answer: In this case I am role-playing as a gameplay engineer candidate applying to Naughty Dog sometime following the release of their game The Last of Us, Part II.

- I really enjoyed how the game narratively explored the perils of war and how violence can lead to a viscous cycle of revenge. My favorite aspect of this was when I began to feel that story expressed through the gameplay.
- For example, when I played as the character Ellie, I had to fight some people that I had made friends with when I had earlier played as the character Abby. That was emotionally challenging as a player. I wish the game provided me with more opportunities

to explore taking a more pacifist route when I began to feel uncertain of using violence.

- For example, and this was a pain-point I encountered from The Last of Us Part I as well, I would sometimes be in an encounter where I felt it would make the most sense for Ellie to thematically just run away... yet the encounter required me to kill all enemies in the vicinity before I could progress. Once I was stuck in an arena and tried to escape through what looked like an exit, but it did not work. Then I eventually fought the arena's enemies to discover that the exit I was trying to use would only prompt me with an escape action once I had killed all the people nearby.

- So, if I were to think through improving that experience, I would have allowed the player to use the exit before all enemies were killed. I can understand that sometimes for story-reasons this may not be desired. In those cases, I would still want to provide the player with some sort of indicator to tell them that they did indeed find the exit... but that it is not currently available because killing enemies is a requisite objective. We might achieve that through non-diegetic UI, such as an interact prompt that is visible but faded or crossed out to indicate that it cannot be engaged. We could also literally display the text "Cannot use while enemies are nearby". Lastly, we could diegetically relay this information such as having Ellie whisper to herself, "I think I need to deal with the raiders before I leave".

In my answer, I complemented their game and thought through one of its core objectives. Then I explained a moment when the game did not reach its full potential in pursuit of that objective and explored potential solutions to improve it.

For the rest of the problems, we are going to take a tour through some gameplay engineering skills as we analyze a bug in the imaginary game, Money Business. Remember that some of these problems have several solutions. With each problem, think through your options and explain the trade-offs of your approach.

Practice Problem 14.15: You are hard at work developing the AAA game, Monkey Business. A problem has come to your attention concerning the main NPC AI type of the game, the "monkey". You notice that levels which include monkeys are running slower than others. More specifically, the game is running at a lower frame rate during these levels. How would you approach debugging this AI?

Hint: Think through the tools that you would likely have at your disposal. What suspicions do you have with respect to the source of the slowdown? How would you be able to confirm or reject those hypothetical causes?

Solution: Good answers should convey familiarity and confidence with modern debugging tools. Good candidates will already have reasonable suspicions about what could be the issue, and they will explain how they can use their tools to narrow down the potential causes of the problem. They might ask insightful clarifying questions such as "does the issue seem to be affected by the number of monkeys in a level?"

Example Answer: I'd first want to confirm that it is indeed the monkey AI that is causing the slowdown. Using debug tools, I would quickly play the level without the monkey and observe if frame rate restored to a normal number. I'd like to see if there is anything the monkey is doing in particular that might stand out as particularly expensive. I'd open the game's frame debugger to see if it is allocating a large, or very frequent, amount of memory. I'd enable any debug logging to see if an error is being reported. I'd also be curious why this bug is being reported now, was this always the case for the monkey AI or could something have changed recently?

Practice Problem 14.16: Thanks to your analysis, you gain further clarity on the situation. The monkeys seem to slow down the level, specifically when they are targeting other AI to do their special banana attack. During the targeting process of the banana attack, the monkey needs to first ensure that its target is not next to any walls, and this is an expensive computation. So, let's say you have about five monkeys and every frame each of them is going through the banana attack targeting process. What are some approaches you might take toward optimizing this procedure?

Hint: A serious mistake has been made with respect to handling the AI's computation, can you identify the issue? Consider the frequency of the attack validation procedure.

Hint: <u>Every</u> monkey AI is performing the targeting process <u>every</u> frame, why is that wrong? How could it be fixed?

Solution: Candidates should readily identify the glaring mistake in this AI design. The monkeys are processing their attack logic on every frame update. Frame updates correspond to the frequency at which the screen visuals should be updated, an AI's "think" should always be divorced from this graphical update loop. In other words, even if the game runs the graphics update loop at 60 FPS, the monkey does not need to validate its attack sixty times every second.

We should find a more appropriate update interval. For example, it may be sufficient for the monkey to only check if its target can be attacked every two seconds. In addition to changing the AI think frequency, we can round-robin the AI so that they are not all thinking on the same frame. We can also cache the result so that this calculation is not repeated unless the target moves, or the attack is otherwise invalidated. If multiple monkeys are all just checking the player's position maybe they could share this data and only do one calculation, then share the result.

Candidates may also suggest that five AI attacking simultaneously is a potential issue. Of course, that depends on the game design, but keen candidates might suggest using systems like AI attack tokens to stagger the monkeys' attacks. Cooldowns can also be used to prevent overly frequent calculations. We can also spread, aka amortize, the calculation itself over several frames if it's possible to break the calculation into steps.

Example Answer: I would divorce the AI think loop from the game's graphical update loop. Then potentially introduce caching, cooldowns, amortization, tokens, round-robin evaluation, and shared data (blackboarding) to reduce the calculation's performance impact.

Practice Problem 14.17: Thanks to your fixes, the banana attack targeting calculation is now sufficiently fast. However, that has turned your attention to another issue. The banana projectile has some very expensive graphics and display logic. Once the bananas are spawned, they slow down the game. In fact, the game can only handle maybe a dozen of them before it begins to lag regardless of if they are on-screen. You've consulted with the graphics team, and they have determined that the expensive effects are necessary and that the shader itself cannot be optimized. At a high-level what are some approaches you might take to optimizing this problem?

Hint: Something important is not happening if the graphics are still incurring expense despite being off-screen.

Hint: How can you reduce the cumulative expense of several bananas' graphics without compromising their individual visual fidelity?

Solution: There is a clear problem if the graphics are incurring a large expense despite not being shown. When graphics are off-screen or occluded, they should be culled. Candidates should reference both frustum and occlusion culling which may be relevant to this issue. Concerning the number of bananas, we can explore trade-off decisions like preventing banana attacks if several bananas are already on-screen, or using level-of-detail scaling which may be applied to bananas if they are far enough away to not compromise visual fidelity. While some candidates may be drawn to pooling solutions, the problem statement did not indicate that banana instantiation was the main issue.

Example Answer:

If the graphics are very expensive despite not being shown, then something has gone wrong with the game's culling procedure. We should ensure that bananas which are not visible, due to obstructions, are occlusion culled. Likewise, bananas beyond the player's field-of-view should be frustum culled. If some bananas are visible but very far, we may be able to optimize their display without sacrificing fidelity with approaches such as billboarding and LOD scaling.

Practice Problem 14.18: You notice that once the projectiles have spawned the game is performant. But at the exact time of each spawn, the game hits a serious lag spike. What is the problem occurring here and how can we fix it?

Hint: Spawning objects can require an expensive runtime allocation. How can we anticipate the spawn to prevent the mid-game lag spike?

Solution: The expensive spawn is probably causing some memory allocation at runtime. Usually with videogames you want to avoid this issue by pre-loading any allocations you will require throughout the course of the program and then doling out those objects over time, returning them when expired. This paradigm is called pooling.

Example Answer:

I would work with designers to determine the maximum number of bananas we want to support during gameplay. I would then

implement a pool that spawns exactly that many bananas during the load of the level, activates them only when needed, and
returns expired bananas to the pool. If the designers cannot toler-
ate a max bananas count, then we can still pre-load our expected
number of bananas and dynamically allocate any bananas that
exceed our current pool size.

Practice Problem 14.19: You decide to implement a basic pooling system for the
bananas. Should you use an array or a linked list for storing the pool's elements?
What factors informed your decision?

Hint: Arrays, unlike linked lists, store their memory contiguously. What
benefits does that design provide?

Solution: An array is a set amount of contiguous memory, so if we need
to grow its size that reallocation can be expensive. But in the case of most
pool designs, where we have a known max, a resize should never happen.
The array's contiguous memory provides cache coherency when iterating
through the entries. On the other hand, a linked list could spread out its data
anywhere which could make traversing the object list of the pool expensive. Additionally, the linked list's pointers add to the amount of memory
needed. The array is probably a better choice here since we have a known
maximum and we're really just holding the elements rather than doing any
special operations with the data.

Practice Problem 14.20: You create a basic pooling system that uses a shared
global banana array to speed up banana spawning for the monkey AI. Monkeys can
now request and release bananas from the pool using separate threads. But soon
after committing your code, a strange bug is reported. It seems that when more
than one monkey uses this pool, the bananas seem to exhibit odd behavior when
spawned. This behavior includes them teleporting. The more monkeys, the more
likely this behavior occurs. What could be going on here and how can we fix it?

Hint: If every monkey is on a separate thread, what could happen when
they try to edit a shared resource?

Hint: How can the shared resource be secured from race conditions and concurrent access issues?

Solution: If multiple AI threads are sharing the global banana pool, there is probably some sort of race condition or data corruption happening between threads. We don't know exactly how these AI are implemented but we should probably suspect a problem like two AI receiving a reference to the same banana and there being a write collision. If the pool does not already have multithreaded support, there are a few approaches we could use to add that support.

We could add a simple lock. A mutex for the entire pool would stall requests for a banana until the previous request has been processed. This would cause contention, aka when one thread is waiting for another, which is less than ideal.

Another approach is to use atomic variables or atomic functions, such as in a lockless list (that uses CAS) to remove the banana from the pool.

Another approach would be to convert the banana spawn into a request to a managed pool. The pool can maintain a list of queued requests and handle them all in bulk at a specific point in the frame. Then the monkey can be somehow notified when the bananas they requested have been spawned. This design does not rely on the banana being available on the same frame as the request, so you have time to process an expensive allocation if you want to grow the pool.

Practice Problem 14.21: You decide to redesign the banana pooling using a linked list storage system with multithread access supported with CAS. Each monkey thread can now request and release bananas, and it seems to be working well. But the game, very rarely, is leaking memory. You designed it so that each linked list node has a next-pointer. When a banana needs to be removed from the pool, the list is traversed along the next-pointers until we reach the node that we want. Then we use CAS to change the next-pointer before our desired node to instead point to the node after our desired node. What's wrong with our design here? How can we fix it?

Hint: Let's walk through a hypothetical scenario…

- Consider two hypothetical monkey threads, #1 and #2 that are accessing a linked list of elements that look like this: START → A → C → END.

- Thread #1 finds A and decides to remove it from the list. It notes that C follows A so it prepares to complete a CAS that will make START point to C instead of A.
- Before thread #1 executes the CAS, thread #2 inserts a new node, B, in between A and C. This results in the linked list looking like this: START → A → B → C → END.
- Now thread #1 continues with its work to remove A and it completes the work since the CAS will be approved because START still points at A. The final result (with A now removed) is START → C → END, with B dangling. B has leaked! How can we avoid this?

Solution: This is the classic ABA problem that can occur with CAS deletion. Candidates should recognize that a single CAS is never going to be enough on its own to implement a safe deletion. We need to augment the CAS to prevent this problem. There are many ways to fix this. A simple approach is for the deleting thread to make two passes through the list. First to just mark the node as deleted, and second to remove the node that has been marked as deleted. You could add some sort of simple atomic markup on the node to identify its state/ownership. There are other operations available on certain hardware such as LL/SC (linked-load/store-conditional) which will fail if memory has been touched prior to the store operation.

You fix the banana code and Monkey Business is released to worldwide fanfare and massive critical success. Congratulations, you're a game developer.

WORKS CITED

[1] S. Meyers, "code::dive conference 2014 - Scott Meyers: CPU Caches and Why You Care," *YouTube*, 2014. [Online]. Available: https://www.youtube.com/watch?v=WDIkqP4JbkE. [Accessed: Aug. 25, 2024].

[2] "Fox Roxy Toy," *TurboSquid*, 3D model, 2023. [Online]. Available: https://www.turbosquid.com/3d-models/fox-roxy-toy-3d-model-1507450. [Accessed: Aug. 25, 2024].

[3] Frebers, "Gamepad," *Noun Project*, 2020. [Online]. Available: https://thenounproject.com/icon/gamepad-7137431/. [Accessed: Aug. 25, 2024]. License: CC BY 3.0.

Index